Agricultural Drainage and Water Quality

Agricultural drainage is the removal of excess water from the soil surface and/or soil profile of cropland, by either gravity or artificial means. The two main reasons for improving the drainage on agricultural land are for soil conservation and enhancing crop production.

The management goal of agricultural drainage is to maintain a salt balance in the crop root zone in arid areas and proper soil water balance in humid areas. Drainage water from different locations and/or facilities will have different quality characteristics. Poor quality water should be separated from good quality water. If drainage water is unsuitable for reuse, it should be disposed of in a sink of lower quality water.

In recent years, public concern has increased about the nature of agricultural drainage and the impact of agricultural drainage improvements on the quality of water resources and environment. This publication is designed to help citizens to understand the purpose and nature of agricultural drainage systems and water quality issues. It will be a highly useful reference tool for farm and project managers, consultants, engineers and students and researchers in the field of agricultural science.

Premjit Sharma is having more than fifteen years of teaching and research experience mainly on cropping systems and soil fertility. Throughout his career, he has been associated with a number of educational and research institutes. He has published a number of research articles in reputed journals. His areas of interest include agriculture, agricultural economics, agricultural business management, natural resource management and sustainable development.

He has been associated with a dozen of UG and PG Courses in the discipline of agriculture and is the chairman of advisory committee of M.Sc. and Ph.D. students. Apart from being an active and dedicated teacher, he has also been committed in research and running projects funded by the eminent bodies.

Agricultural Drainage and Water Quality

Editor
Premjit Sharma

2018
Gene-Tech Books
New Delhi - 110 002

ISBN 81-89729-26-8

Published by : **GENE-TECH BOOKS**
4762-63/23, Ansari Road, Darya Ganj,
NEW DELHI - 110 002
Ph.: 23262132
e-mail: genetechbooks@yahoo.co.in

Printed at : **Tarun Offset Printers**
Delhi

PRINTED IN INDIA

Preface

Agricultural drainage is the removal of excess water from the soil surface and/or soil profile of cropland, by either gravity or artificial means. The two main reasons for improving the drainage on agricultural land are for soil conservation and enhancing crop production. Most agricultural producers improve the drainage on their land to help create a healthier environment for plant growth and to provide drier field conditions so farm equipment can access the farm field throughout the crop production season. Healthy, productive plants have the potential to produce greater yields and more food. Also, agricultural drainage improvement can help reduce the year-to-year variability in crop yield, which helps reduce the risks associated with the production of abundant, high quality, affordable food.

The management goal of agricultural drainage is to maintain a salt balance in the crop root zone in arid areas and proper soil water balance in humid areas. Drainage water from different locations and/or facilities will have different quality characteristics. Poor quality water should be separated from good quality water. If drainage water is unsuitable for re-use, it should be disposed of in a sink of lower quality water. Highly saline subsurface drainage water from arid areas, re-used for irrigation, could affect the growth of salt sensitive crops. In humid areas, most subsurface drainage water has the potential to be reused. There are several factors to consider when determining the constraints for the management of surface or subsurface agricultural drainage water. The amount and quality of drainage water managed, changes in the rate of flow, and chemical concentrations need to be determined.

In recent years, public concern has increased about the nature of agricultural drainage, and the impact of agricultural drainage improvements on the quality of water resources and environment. This publication is designed to help citizens to understand the purpose and nature of agricultural drainage systems and water quality issues. It will be a highly useful reference tool for farm and project managers, consultants, engineers and students and researchers in the field of agricultural science.

Editor

Contents

1

Introduction

Drainage is an important water management practice, and it is a component of efficient crop production systems. World food supply and the productivity of existing agricultural lands can only be maintained and enhanced if drainage improvements are undertaken on cropland currently affected by excess water and high water tables. Drainage (both surface and subsurface) is not simply the conversion of wetlands, but the improvement of naturally inadequately drained cropland. It is complementary to irrigation and is viewed as an essential component of irrigated agriculture. The objective is to increase production efficiency, crop yields and profitability on naturally poorly drained agricultural lands.

Excess water in the crop root zone soil is injurious to plant growth. Crop yields are drastically reduced on poorly drained soils, and, in cases of prolonged waterlogging, plants eventually die due to a lack of oxygen in the root zone. Sources of excess soil water that result in high water tables include: high precipitation in humid regions; surplus irrigation water and canal seepage in the irrigated lands; and artesian pressure. Waterlogging in irrigated regions may result in excess soil salinity, i.e., the accumulation of salts in the plant root zone. Artificial drainage is essential on poorly drained agricultural fields to provide optimum air and salt environments in the root zone.

The primary benefits of drainage go beyond the control of excess soil water and accumulation of excess salts in the crop root zone. The coincident environmental and socio-economic benefits associated with disease vector control and public health must be fully recognised. One of the major environmental benefits of drainage is its positive impact on improving the health of humans, plants and farm animals. Drainage of wet, swampy areas has led to a reduction in mosquito breeding sites in all parts of the world. The effect has been a drop in the incidence and prevalence of important water related and mosquito transmitted diseases, e.g., malaria, yellow fever and filariasis.

Furthermore, drainage of stagnant water can eliminate foot-rot in large animals and, to a certain extent, the breeding environment of aquatic and semi-aquatic snails, which are the intermediate host of schistosomiasis. Drainage also reduces or eliminates mildew infections and various root rots of plants. The overall impact of improved drainage has been an improvement in hygienic conditions, in the health sector and in the productivity of human beings. By growing high value food crops in well-drained soils, the health, nutrition and economic status of rural populations can be improved.

There are also increased opportunities for employment, as new industries may develop in prosperous areas. Where drainage is used to reclaim salinised and waterlogged lands, it is an environmentally beneficial practice, because the land is returned to its full productive potential. The adaptation of subsurface drainage systems to serve as sub-irrigation or controlled drainage systems leads to other benefits, i.e., the reduction of nitrate pollution. The field-scale benefits of drainage are:

— Drainage promotes beneficial soil bacteria activity and improves soil tilth.

— There is less surface runoff and soil erosion on drained land.

— Improved field machine trafficability reduces soil structural damage. Soil compaction is reduced and less energy is required for field machine operations. Drainage also allows for more timely field operations. Consequently, the growing season can be lengthened and crops can achieve full maturity.

— Crop yields are increased because of improved water management and uptake of plant nutrients.

— Higher value crops can be planted, and there is flexibility to introduce new and improved cropping systems.

— In general, land value and productivity are increased.

— Farm income is increased and income variability reduced.

— Drainage maintains favourable salt and air environments in the crop root zone.

During the late 1800s, European settlers in the Upper Midwest began making drainage ditches and channelising (straightening and reshaping) streams to carry water from the wet areas of their farms to nearby streams and rivers. Later, farmers increased drainage by installing subsurface drainage pipes generally at a depth of three to six feet.

Until the 1970s, most subsurface drainage pipes were made from short, cylindrical sections of concrete or clay called "tile." That is why terms like tile, tile drainage, and tiling are still used, even though most drainage pipe today is perforated polyethylene tubing. When installing a subsurface drainage system, pipes are either strategically placed in a field to remove water from isolated wet areas or installed in a pattern to drain an entire field. In some areas, surface inlets or intakes (risers extended from underground pipes to the surface) remove excess surface water from low spots in fields.

Many soils in the Upper Midwest, as well as soils in other regions of the U.S. and the world, have poor natural internal drainage and would remain waterlogged for several days after excess rain without artificial drainage. This prolonged wetness prevents timely fieldwork and causes stress to growing crops because saturated soils do not provide sufficient aeration for crop root development. The roots of most crops grown in Minnesota cannot tolerate excessively wet conditions for more than a couple of days.

Soil conditions that make drainage a necessity for some agricultural lands include those with slow water permeability or dense soil layers that restrict water movement, flat or depressional topography and, in some areas, high levels of salts at the soil surface. Large areas of Minnesota would not reliably produce crops if artificial drainage systems had not been installed. Subsurface drainage pipes are typically placed at depths of 3 to 4 feet in poorly drained soils.

Farmers must make a significant financial investment when installing an agricultural drainage system. They are willing to make this investment for two major reasons:

1. Agricultural drainage systems usually increase crop yields on poorly drained soils by providing a better environment for plants to grow, especially in wet years.
2. The systems generally help improve field conditions for timely tillage, planting and harvesting.

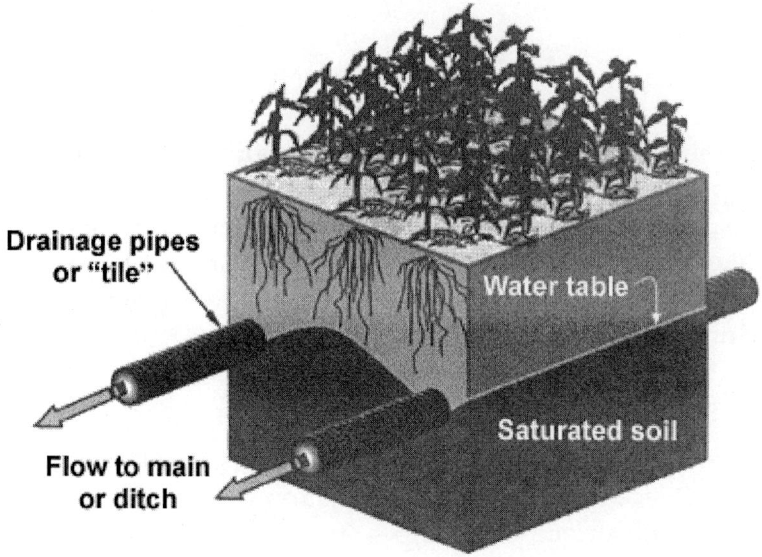

The most recent USDA comprehensive survey of drained lands showed that in 1985, 30 percent of all agricultural lands in the Upper Midwest (Illinois, Indiana, Iowa, Michigan, Minnesota, Missouri, Ohio, and Wisconsin) were artificially drained. Minnesota has large areas of poorly drained soils: e.g., 66 and 59 percent of the soils in the Red River and Minnesota River basins, respectively. In recent years, farmers have installed as much as 100 million feet of subsurface drainage pipe in Minnesota annually.

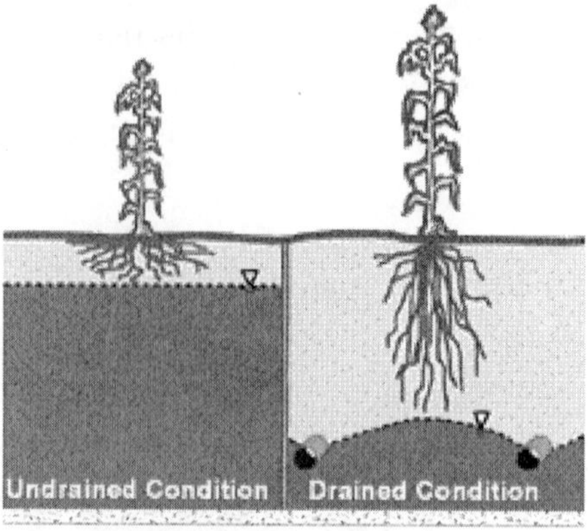

A significant portion of new drainage activities is replacement and enhancement of old drainage systems. As old systems age and decay, replacement activities will likely continue. Subsurface drainage promotes better root growth and plant health when soils have poor internal drainage.

Different Types of Drainage Systems

Agricultural land drainage usually consists of surface or subsurface systems, or a combination of both. At the field scale, subsurface drain pipes and field ditches normally exit to an open main or collector drain. At the regional level, the latter then empties into a river or its tributaries. In some instances, depending on the character of the hydrological basin, main drains may dispose of drainage water to an evaporation pond, to a wetland, or to a saline agriculture/ agriculture-forestry system.

Surface Drainage

Surface drainage is often achieved by land forming and smoothing to remove isolated depressions, or by constructing parallel ditches. Ditches and furrow bottoms are gently graded and discharge into main drains at the field boundary. Although the ditches or furrows are intended primarily to convey excess surface runoff, there is some seepage through the soil to the ditches, depending on the water table position. This could be regarded as a form of shallow subsurface drainage. Surface drainage is especially important in humid regions on flat lands with limited hydraulic gradients to nearby rivers or other disposal points. There is also a need for good surface drainage in semi-arid regions which are affected by monsoons.

Subsurface Drainage

Surface drainage alone is seldom sufficient to remove excess water from the crop root zone. Deep ditches or subsurface pipe drainage systems enable a more rapid water table drawdown. The downstream ends of the laterals are normally connected to a collector drain. The required diameter of the pipe collectors increases with the area drained. Drain spacing is usually dependent on soil hydraulic conductivity and a design drainage rate coefficient. Depending on topography, land formation and proximity of a water receiving body, the collector may outlet by gravity to an open main drain or into a sump.

In the latter case, the discharge is then pumped to another drain, or ultimately to a lake or stream. In some flatter parts of eastern Canada, the eastern and mid-western United States, and parts of Europe, subsurface pipe drains are also used for sub-irrigation. In this case, in dry periods, surface water is introduced into the drain pipe system from an external source, and the water table is raised. Moisture then moves upward by capillary action to the root zone. Sub-irrigation is regarded as a highly energy and water efficient method of irrigation. In another process, known as controlled drainage, elevated water tables can be maintained with a control structure on the collector pipe. Horizontal subsurface drainage systems are used in irrigated arid and semi-arid regions to reclaim saline and waterlogged lands, and to maintain favourable long-term salt and water balances in the crop root zone.

Salinity and waterlogging are caused by a build up of the water table due to deep percolation of normal excess water and canal seepage. Buried pipe drains are generally installed deeper in arid regions than in humid regions in order to control salinity. Water in excess of plant evapotranspiration (ET) needs is always unavoidably applied during irrigation. This additional quantity of water applied is known as the leaching fraction. Naturally occurring as well as applied salts are then leached from the root zone by this water, and removed from the field via the pipe drains. Deeper drain installation ensures that salts do not rise too rapidly to the soil surface due to capillary action.

Drainage also prevents waterlogging of the root zone. The amount of irrigation water to be removed is generally less in arid than in humid regions. Vertical drainage by means of tube-wells is also used to control waterlogging and salinity in some parts of the world, e.g., India, Pakistan and central Asian republics. The primary purposes of tube-wells are the same as those of horizontal drains, and at the same time to extract groundwater for irrigation. As a result of pumping, the water table is lowered, and salinisation due to capillarity is minimised. This situation is ideal where the groundwater is not very brackish or saline, and is therefore suitable for irrigation. In areas where the groundwater is highly saline, the pumped water may be too saline for irrigation, unless mixed with fresher or less saline water. Where the groundwater is too saline for crop production, it must be disposed of. Drainage does not have a direct impact on groundwater quality. It only serves to collect and transport excess water.

Secondary Drainage Treatments

Methods of improving the internal drainage of low permeability soils include: subsoiling, deep tillage, mole drainage, and biological practices, viz., cropping with deep rooted legumes (e.g., alfalfa) and crop rotations. In some parts of the world, deep rooted trees are used to lower the water table. There are usually no water quality hazards associated with these supplemental drainage practices.

Artificial Drainage

Wetland soils may need drainage to be used for agriculture. In the northern USA and Europe, glaciation created numerous small lakes which gradually filled with humus to make marshes. Some of these were drained using open ditches and trenches to make mucklands, which are primarily used for high value crops such as vegetables. The largest project of this type in the world has been in process for centuries in the Netherlands. The area between Amsterdam, Haarlem and Leiden was, in prehistoric times swampland and small lakes. Turf cutting (Peat mining), subsidence and shoreline erosion gradually caused the formation of one large lake, the Haarlemmermeer, or lake of Haarlem.

The invention of wind powered pumping engines in the 15th century permitted drainage of some of the marginal land, but the final drainage of the lake had to await the design of large, steam powered pumps and agreements between regional authorities. The elimination of the lake occurred between 1849 and 1852, creating thousands of km^2 of new land. Coastal plains and river deltas may have seasonally or permanently high water tables and must have drainage improvements if they are to be used for agriculture. An example is the flatwoods citrus-growing region of Florida.

After periods of high rainfall, drainage pumps are employed to prevent damage to the citrus groves from overly wet soils. Rice production requires complete control of water, as fields need to be flooded or drained at different stages of the crop cycle. The Netherlands has also led the way in this type of drainage, not only to drain lowland along the shore, but actually pushing back the sea until the original nation has been greatly enlarged.

In moist climates, soils may be adequate for cropping with the exception that they become waterlogged for brief periods each year, from snow melt or from heavy rains. Soils that are predominantly clay will pass water very slowly downward, meanwhile plant roots suffocate because the excessive water around the roots eliminates air movement through the soil. Other soils may have an impervious layer of mineralised soil, called a hardpan or relatively impervious rock layers may underlie shallow soils.

Drainage is especially important in tree fruit production. Soils that are otherwise excellent may be waterlogged for a week of the year, which is sufficient to kill fruit trees and cost the productivity of the land until replacements can be established. In each of these cases appropriate drainage is used to carry off temporary flushes of water to prevent damage to annual or perennial crops.

In farming drier areas, irrigation is often used, and one would not consider drainage necessary. However, irrigation water always contains minerals and salts, and these can be concentrated to toxic levels by evapotranspiration. Irrigated land may need periodic flushes with excessive irrigation water and drainage to remove these toxic minerals.

Modern Drainage

Modern drainage systems incorporate geotextile filters that retain and prevent fine particles of soil from passing into and clogging the drain. Geotextiles are synthetic textile fabrics specially manufactured for civil and environmental engineering applications. Geotextiles are designed to retain fine soil particles while allowing water to pass through. In a typical drainage system they would be laid along a trench which would then be filled with coarse material: gravel, sea shells, stone or rock. The geotextile is then folded over the top of the stone and the trench is then covered by soil. Groundwater seeps through the geotextile and flows within the stone to an outfall.

In high groundwater conditions a perforated plastic (PVC or PE) pipe is laid along the base of the drain to increase the volume of water transported in the drain. Alternatively prefabricated plastic drainage systems, often incorporating geotextile, coco fiber or rag filters can be considered. The use of these materials has become increasingly more common due to their ease of use which eliminates the need for transporting and laying stone drainage aggregate which is invariably more expensive than a synthetic drain.

Over the past 30 years geotextile and PVC filters have become the most commonly used soil filter media. They are cheap to produce and easy to lay, with factory controlled properties that ensure long term filtration performance even in fine silty soil conditions.

Tile Drainage

Tile drainage is a practice for removing excess water from the subsurface of soil intended for agriculture. Whereas irrigation is the practice of adding additional water when the soil is naturally too dry, drainage brings soil moisture levels down levels for optimal crop growth. While surface water can be drained via pumping and/or open ditches, tile drainage is often the best recourse for subsurface water. Too much subsurface water can be counterproductive to agriculture by limiting access to the land, particularly by farm machinery, and inhibiting the growth of crops.

In terms of access, most modern agriculture depends on the usage of large machinery—tractors and implements—to prepare the seedbed, plant the crop, carry out any cultivation and applications during the growing season, and ultimately, to harvest the crop. Operating most machinery in excessively wet conditions may result in soil degradation and inhibit the operation of the machinery (ie, "getting stuck"). Most cash crops require specific soil moisture conditions, and do not grow well in wet, "mucky" soil.

In a tile drainage system, a sort of "plumbing" is installed below the surface of agricultural fields, effectively consisting of a network of below-ground pipes that allow subsurface water to move out from between soil particles and into the tile line. Water flowing through tile lines is often ultimately deposited into surface water points—lakes, streams, and rivers—located at a lower elevation than the source. Water enters the tile line either via the gaps between tile sections, in the case of older tile designs, or through small perforations in modern plastic tile. Soil type greatly affects the efficacy of tile systems, and dictates the extent to which the area must be tiled in order to ensure sufficient drainage. Sandier soils will need little, if any, additional drainage, whereas soils with high clay contents will hold their water tighter, requiring tile lines to be placed closer together.

Throughout the twentieth century, the technology of tile installation remained similar to the methods first used in 1838. Although cement sections later replaced the original clay tiles, and machines were used to dig the trenches for the tile lines, the process remained quite labour-intensive and limited to specialised contractors. The introduction of plastic tile served to reduce both the cost of tile installation, as well as the amount of labour involved.

Rather than set individual sections of cement tile end-to-end in the trench, tile installers had only to unroll a continuous section of lightweight, flexible tile line. Towards the end of the twentieth century, when large, four-wheel-drive tractors became more common on American farms, do-it-yourself tile implements appeared on the market. By making tile installation cheaper and allowing it to be done on the landowner's schedule, farmers are capable of draining localised wet spots that may not create enough of a problem to merit more costly operations. In this way, farmers may enjoy slight increases in crop yield.

Effects of tile drainage

Unfortunately, the ability for farmers to install their own tile can be problematic. First of all, private installations may reduce the ability of local drainage supervisory boards to regulate tile installation, which in some areas of the country requires proper documentation before a contractor can continue. This leads to the second potential conflict, the unintentional interruption of existing tile networks. Most "do-it-yourself" tile plows do not dig trenches but rather split the soil enough to squeeze the tile line in; thus, a farmer would not be aware if he breaks a line of tile that might serve his neighbours, as well (Mutual tile lines are often dictated by topography rather than land ownership, and the location of many old, but still effective, tile lines are unknown.).

The potential for across-the-fence disputes are obvious. Ecologically, the expansion of drainage systems has had tremendous negative effects. Hundreds of thousands of wetland species experienced significant population declines as their habitat was increasingly fragmented and destroyed. Although market hunting within the Central Flyway was a contributing factor in the decline of many waterfowl species' numbers in the early decades of the twentieth century, loss of breeding habitat to agricultural expansion is certainly the most significant.

Early maps of midwestern states depict many lakes and marshes that are either nonexistant or significantly reduced in area today. Channelisation, a related process of concentrating and faciliting the flow of water from agricultural areas, also contributed to this degradation. In bypassing the natural flow of water from the surface to the water table, drainage systems often prevent the natural filtration of water provided by soils and wetlands. Thus, drainage systems pose a threat to surface water sources by directly depositing water laden with fertilisers, eroded soil, agrochemicals, and other types of agricultural runoff pollutants. In very flat areas, where the natural topography does not provide the gradient necessary for water flow, "agricultural wells" can be dug to provide tile lines sufficient outlet. In these cases, it is the groundwater that stands to be polluted by unfiltered tile output.

Today, a number of state and federal initiatives serve to reverse habitat loss. Many programmes encourage and even reimburse farmers for interrupting the drainage of localised wetholes on their property, often by breaking tile intakes or removing the tile completely. Landowners are often partially or fully compensated for forfeiting the ability to grow crops on this land. Such programmes and the cooperation of landowners across the country have had significant positive effects on the populations of a wide variety of waterfowl.

Management of Drainage Water

Drainage water can be disposed directly to open surface water bodies, e.g., rivers, lakes, outfall drains, seas or oceans. Wetlands, evaporation ponds and solar evaporators are used as discharge points where there is no direct outlet to one of the open surface water bodies mentioned above. This is generally the situation in hydrologically closed drainage basins. Some re-use of drainage water for irrigation is another management option, where sequential irrigation of increasingly salt-tolerant plants, ending with Salicornia, occurs.

Drainage water can also be disposed of by injection to the deep groundwater. If drainage water is disposed of to large, open surface water systems with significant dilution or assimilative capacity, then water quality problems are minimised. However, water quality problems may develop with repeated re-use, disposal in closed basins and injection to deep wells. In this last case, the concern is that if the drainage water contains sufficient amounts of salts, nitrates, bacteria and trace elements, the groundwater could become contaminated. The problem may be especially severe if the aquifer is also used for drinking water supplies downslope.

On-farm Source Control

The most efficient method of minimising environmental problems is to implement source control practices at the farm level. In irrigated areas, this can be achieved by improved irrigation water management. Higher irrigation efficiencies and lined irrigation conveyance structures will reduce the amount of drainage water which needs to be removed. Furthermore, timely and efficient applications of fertilisers and pesticides, used only when necessary, will reduce chemical leaching. Research by Madramootoo et al. has shown that intercropping corn with a legume or ryegrass reduces the amount of nitrates found in drainage water.

Re-use of Drainage Water

In many regions where irrigation water is scarce, drainage water is used to meet crop water requirements. Re-use is only sustainable if the drainage water is of sufficiently good quality. Some of the water quality concerns about drainage water re-use with plants of increasing salt tolerance are that: the effluent may be high in salt content (in irrigated lands); the drainage water can be contaminated with trace elements, toxic organic substances, industrial waste and municipal waste in open main drains. Contaminated drainage water could lead to various problems including: impairment of soil physical and chemical properties, water related health problems, and possible contamination of food products.

Closed Drainage Basins

Ultimate disposal of drainage water to a river or sea is

not always possible. Closed drainage basins present a unique environmental or water quality challenge. In such situations, evaporation ponds may be an appropriate means for disposing of drainage water. However, these ponds may eventually lead to other environmental problems. For example, toxic substances could accumulate in the ponds. Furthermore, in arid climates, as pure water evaporates from the pond, the concentration of the remaining water approaches that of brine.

The health of waterfowl, fish and other aquatic biota which use the pond could be negatively affected. Other environmental problems associated with evaporation ponds include: use of the pond to collect wastewater from homes; human health problems caused by consuming water from the ponds; and the need to ultimately dispose of the accumulated concentrated chemicals in the ponds. In some cases, dry toxic materials may be spread by the wind. Furthermore, these ponds could become habitats for snails and mosquitoes, thereby causing malaria and schistosomiasis epidemics.

In addition, if not properly managed, new waterlogged and saline areas will develop adjacent to the ponds. There is now increasing interest in the utilisation of natural and constructed wetlands to manage drainage water. Wetlands are particularly effective for removing sediment, N and P. Plant, soil and hydrologic parameters interact in a complex way to filter and trap pollutants, and to recycle nutrients. Certain tree and plant species have the potential to absorb pollutants. Residence time, flow rate, hydraulic roughness and wetland size and shape are some of the factors which influence treatment efficiency.

The water supply to the wetland must be sufficient to provide an excess to discharge and prevent salt accumulation. In areas where soils, geologic and hydrologic conditions do not permit constructed wetlands, the saline agriculture/agriculture-forestry system may be appropriate for the disposal of drainage water. The operating principle is to successively re-use saline drainage water to irrigate crops and trees of increasing salt tolerance, and to discharge the final much reduced volume of water into a solar evaporator for salt crystallisation.

The depth of water ponded in the evaporator is regulated to match the daily evaporation rate. The goal is to make the crystallisation pond unattractive to flora and fauna. Another challenge of evaporation ponds is that evaporation is reduced as the ponds become more concentrated. This necessitates the use of additional land to maintain disposal (evaporation) capacity.

Negative Impacts of Drainage

Although agricultural drainage has benefited agricultural production in many regions and countries, there are concerns about its potential environmental impact. Subsurface drainage systems have a positive impact because they generally decrease the amount of surface runoff, thereby reducing the loss of substances generally transported by overland flow. There are concerns, however, about the potential negative impacts of drainage on the hydrology of watersheds, the water quality of receiving water bodies, and the amount and quality of nearby wetlands.

Drainage systems are designed to alter field hydrology (water balance) by removing excess water from waterlogged soils. There are concerns about the downstream hydrological effects caused by draining this excess water. Anecdotal evidence indicates that streams and ditches have become "flashier" over time, spilling over their banks and causing localised crop damage. Some research articles suggest that the most dramatic hydrological changes in a landscape occur when it's converted from native vegetation to agricultural production, and that subsurface drainage may reduce peak flows in some situations.

Surface drainage (enhancing overland runoff) tends to increase the loss of nutrients and sediment that occur with surface runoff. Subsurface drainage, however, can decrease surface runoff, thereby reducing sediment losses by 16 to 65 percent and phosphorus losses by up to 45 percent. The main water quality concern about subsurface drainage is the increased loss of nitrates and other soluble constituents that can move through soil to drainage systems and end up in nearby surface water. In addition, surface intakes, which are common across southern Minnesota and northern Iowa, provide a fairly direct pathway for sediment and other contaminants in surface runoff to reach nearby surface waters.

Despite the fact that wetlands are protected by various regulations, it is estimated that over 60,000 wetland acres are lost nationally each year. The loss of wetland ecosystems—valued for their wildlife habitat, for water storage, and increasingly for their potential role to improve water quality—is not easy to quantify. But it's likely that agricultural and urban drainage activities both cause wetland loss.

Drained Soils Management

Many current drainage research and Extension programmes throughout the country are trying to find ways to reduce the potential environmental impacts of agricultural drainage while retaining its agronomic benefits. Some management practices have been effective; others are presently being examined. Both are described in the following sections:

Improved Nutrient Management

The proper management of crop nutrients (nutrient source, application rate and timing) is an important way to help control the loss of nutrients through surface runoff and subsurface drainage water. It's been shown that the application of nitrogen fertiliser at rates higher than those recommended by the University of Minnesota increases the amount of nitrate removed through subsurface drainage systems. Since university recommendations are based on an optimum economic return, over-application of nitrogen fertiliser should be less profitable. It should be noted, however, that drained agricultural soils have significant nitrate losses from the natural process of organic matter mineralisation. Improved nutrient management can potentially reduce nitrate losses on drained lands by up to 30 percent.

Changes in Cropping Systems

Row crops such as corn and soybeans experience considerably more nitrate loss through subsurface drainage flow than perennials such as alfalfa and brome grass. So the incorporation of alfalfa or other perennials into farmers' crop rotations could significantly decrease nitrate losses to nearby surface water. While alfalfa may be a financially sound crop for some operations, it is not an economically viable solution for many Minnesota farmers.

Design of Drainage System

Subsurface drainage systems are designed to remove excess water from soil quickly enough to minimise crop stress in most years. Agricultural engineers have developed depth and spacing guidelines for installing drainage pipes. For example, recommendations for the many clay-loam soils prevalent in much of southern Minnesota call for placing drainage pipes approximately three feet deep and 60 feet apart or four feet deep and 80 feet apart. Either design should remove water at the same rate and give similar crop yields.

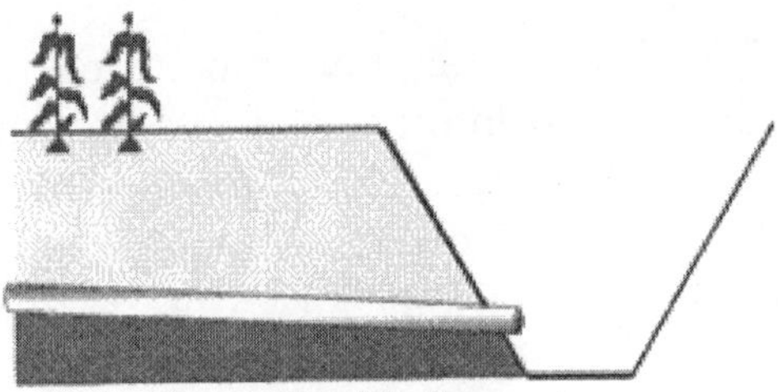

Conventional Drainage

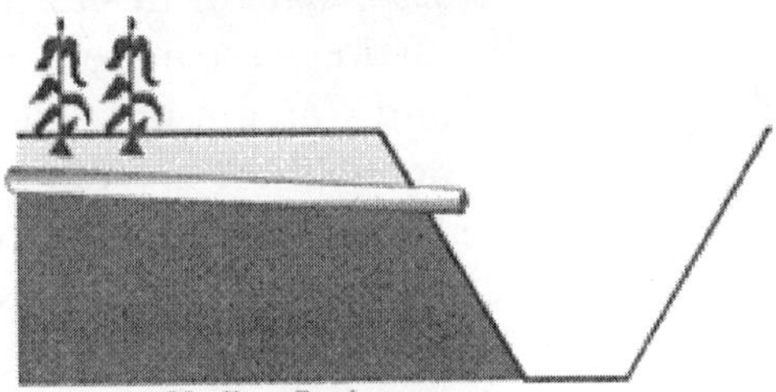

Shallow Drainage

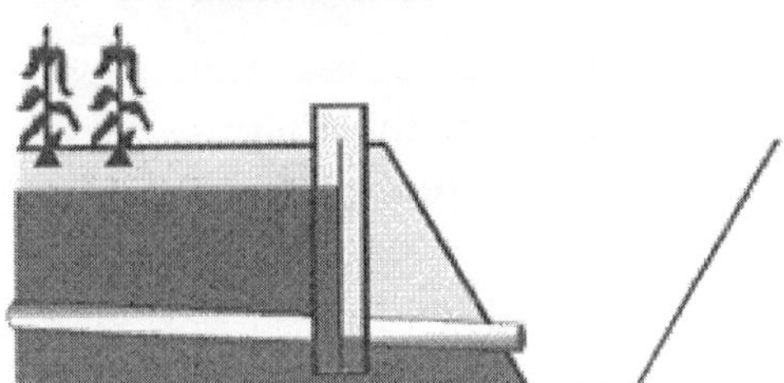

Controlled Drainage

It has been proposed that placing drainage pipes at shallower depths might result in less nitrate loss. This would happen because nitrate would be more likely to reach a biologically active but saturated zone and be converted to nitrogen gas by denitrifying bacteria. The conversion of nitrate/nitrogen to nitrogen gas would prevent the nitrate from reaching the drainage pipes and nearby surface waters. This practice, if proven effective, offers the advantage of being applicable anywhere that drainage systems are installed. It also requires no new management or capital investment.

Controlled Drainage

Controlled drainage has become recognised as an effective practice—and in other states, a best management practice—for mitigating nitrate losses from drainage systems. This practice involves placing simple water control structures at various locations in the system to raise the water elevation. This elevated water causes the water table in the soil to rise, which, in effect, decreases the drained depth of the field. Researchers from North Carolina, Ohio, Michigan and Canada have demonstrated that controlled drainage decreases the volume of water drained (15-35 percent), slightly increases surface runoff (because soils have less space to store water), and significantly decreases (up to 50 percent) nitrate losses seen in conventionally drained fields.

Decreases in nitrate losses have been attributed primarily to reductions in the volume of water drained and, to a somewhat lesser extent, by increased denitrification in the soil. If managed properly, controlled drainage has the potential to improve crop yields by making more water available to plants. The application of controlled drainage techniques is limited, however, by topography. The process is economically unfeasible on land slopes greater than about one percent because more water control structures are needed as slopes increase. In addition, controlled drainage adds new management requirements to systems (also increasing with slope) that some will view as a disadvantage.

Inlet Alternatives

Alternatives to the traditional "open inlet" are being used more frequently around Minnesota. One design involves digging a trench, placing drainage pipe at its bottom, and filling the trench with small rock. These "rock" or "blind" inlets slow the flow of water (compared to open inlets) and may reduce the amount of sediment reaching the drainage system. Another design involves the installation of subsurface drainage pipes in a very tight pattern in a small area in the middle of a wet spot. Another, more traditional, technique involves replacing open inlets with perforated risers. All these designs have the potential to do a better job of protecting water quality than open inlets, while still providing adequate drainage so crops don't "drown."

Wetlands

Wetlands have been proposed as a means of treating water from drainage systems before it is released into nearby rivers or lakes. Biological activity in wetlands can be effective at removing nitrate by converting it to nitrogen gas through a denitrification process that's similar to what occurs in soils. Researchers in Iowa suggest that wetlands can remove from 20 to 80 percent of the annual nitrate in subsurface drainage water depending on the ratio between the areas of drained land and wetland. This approach to "treating" drainage water presents some challenges.

Site topography may pose difficulties in getting subsurface drainage waters to the surface and into wetlands. Land requirements and the cost of construction are also important economic factors. Finally, the bulk of nitrate losses from drained lands in Minnesota occur in early spring when wetlands are not functioning at their peak capacity to remove nitrate, because of low temperatures and high water flow rates. The potential effectiveness of wetlands in treating drainage water in colder climates requires more research.

Environmental Impact Assessment (EIA)

Many institutions require an environmental impact assessment (EIA) prior to construction activities associated with new projects or the rehabilitation of existing projects. The objectives are to identify potential adverse environmental effects, the magnitude of these effects, and to develop mitigative measures. Positive benefits are also identified. In cases where the adverse impacts far outweigh the expected benefits, the project may be completely redesigned or suspended. This is also the case where mitigative measures would be either too costly or technically not feasible. The EIA should be conducted in the earliest stages of decision making, when crucial decisions are still being deliberated. According to Ochs and Bishay, environmental effects can be classified as:

- direct and indirect, or first order and higher order. These are chain effects which are felt throughout, and possibly downstream of a catchment.
- secondary. The primary activity of a drainage project may be extended to include secondary activities.

— synergistic. These effects include an increased threat to the survival of certain species of wildlife that are under pressure in several ways as a result of the same project.

Following are the main steps in the EIA process:

— Scoping. This is a public process, involving the participation of all parties. It results in specific guidelines for inclusion in the environmental impact study (EIS). (EIS refers to the document, whereas EIA refers to the entire process).

— Drawing up the EIS. Potential adverse and beneficial impacts are identified. Proposed actions with alternatives are indicated and mitigative measures are presented.

— Submitting the EIS for public review.

— Receiving advice from the reviewing agency.

— Accepting the EIS.

— Choosing project components. The decision-maker chooses the project option to be implemented and the mitigative measures.

— Implementing the project.

— Monitoring. During and after project implementation, the actual environmental impacts are monitored and compared with the EIS predictions. This is useful for improving future predictions and project designs.

Matrices and checklists are some of the tools most commonly used in an EIA to assess the magnitude or relative weight of both positive and adverse environmental effects. It is hoped that the criteria on drainage water quality provided in the publication will further assist in the evaluation of EIA of irrigation and drainage projects.

The installation of drainage systems may result in changes to the associated ecosystem. These changes may be either beneficial or adverse. However, there are potential adverse water quality impacts associated with drainage. The concentrations of salts, nutrients and other crop-related chemicals in drainage discharge vary with time and discharge rate. The use of fertilisers and pesticides in intensive agricultural production has sometimes led to damage to downstream aquatic ecosystems.

Drainage planners therefore need to analyse effluent for nutrients and pesticides. The nutrients of most concern are N and P. In addition, from time to time, natural trace elements from the soil itself may be harmful to the ecosystem. Effluent laden with N and P stimulates eutrophication in receiving water bodies. In addition to agricultural chemicals and trace elements, drainage water from irrigated areas frequently contains salts. The impact of salts on downstream users needs to be evaluated. Some soils are abundant in trace elements, and these could leach to the drainage system.

Small amounts of trace elements such as As, Cd, Hg, Pb, B, Cr and Se are harmful to aquatic species because of biological magnification. The environmental consequences of disposing of drainage water from California's irrigated San Joaquin Valley into the 470 ha Kesterson Reservoir (a closed basin) are well known. The reservoir was a waterfowl habitat, and concentrating Se in the drainage water caused fish species to disappear, and resulted in deformities in waterbird embryos. The failure to construct an adequate outfall drain to the sea, the use of a waterfowl habitat for drainage water disposal, and the lack of proper water quality monitoring were the major reasons for these negative environmental impacts.

Care must be taken to ensure that the disposal of drainage water does not interfere with the ecosystem's aquatic and terrestrial species. Concern has been expressed about damage by agricultural drainage water to estuarine fisheries in some countries.

2

Surface Drainage

Surface drainage is needed in flatland areas where water floods over the land. Slopes in these areas generally are not great enough for the flowing water to cause erosion; therefore, the object of drainage systems in flatland areas is to remove the excess water before it damages crops. This differs from erosion control work in which the object is to control the erosive velocity of peak flood flows. When improving soils that have very low permeability or reclaiming highly salt affected soils that need drainage, the use of open drains has a practical installation advantage over pipe drainage. At a later stage when the structure of these soils improves and their hydraulic conductivity values become relatively stable, cost effective subsurface pipe drains can be considered.

Drainage Runoff

Ditches for drainage of common field crops generally are designed to remove runoff from the drainage area within a 24-hour period following a rain event. Some surface flooding of the land during this period is permissible. Some high-value and speciality crops require a more rapid rate of removal of runoff to prevent crop damage. For these crops, a 6- to 12-hour removal interval may be necessary during the growing season. The drainage coefficient, is the rate of water removal per unit of area used in drainage design. For surface drainage the coefficient generally is expressed in terms of flow rate per unit of area, which varies with the size of the area.

Curves for determining runoff for drainage design have been prepared for most of the humid areas of the United States. The curves are based on the climate, soils, topography, and agriculture of the particular area. Curves developed by John G. Sutton, known as the SCS drainage runoff curves have been used in the Northern States since the 1930's. The Red River Valley runoff curves, based on the SCS drainage runoff curves and adjusted to the Red River Valley of the North, have been in use for about the same period.

Other drainage curves have been developed for use in the Southeastern, Southwestern, and Western States. Because of the wide variations in rainfall, soils, crops, and topography, drainage design should be based on drainage coefficients applicable to local conditions. The capacity for design given by most of the curves varies in accordance with the general formula (Cypress Creek equation):

$$Q = CM^{0.83}$$

where:

Q = flow in ft^3/s for which the drain is to be designed

C = the appropriate drainage curves

M = area in mi^2 of watershed

The capacity for drainage ditches in arid, irrigated areas is generally determined by the leaching requirement and amount of return flow from irrigation water. If surface runoff from precipitation is involved, the above formula with the appropriate value of C may be used. For many ditches in these areas, the depth rather than the capacity required governs the size of the drain.

Composite Curves

Composite curves are used where runoff from the land draining into an open drain must be based on different drainage curves. The design capacity of ditches in these areas can be determined by computing equivalent drainage area information. Using this method, you determine the runoff from the acreage of one type of land and convert it to the equivalent acreage of another type of land so that the acreage may be added together and used to find the discharge from the area.

Investigation and Planning

A surface drainage system may consist of field ditches and/or land forming with ditches and underground pipes to carry the drainage water to the outlet. The drainage system should provide for an orderly removal of excess water from the surface of the land. Open drains may serve any land use. In urban uses they are typically called open drains or surface storm drains, while in agricultural fields they are usually referred to as ditches. Current legislation and regulations require care to avoid environmental damage to existing wetlands. Maintenance of drainage systems is critical to the sustainability of agriculture and for other land uses.

Field Ditches

Field ditches are shallow ditches for collecting and conveying water within a field. They generally are constructed with flat side slopes for ease in crossing. These ditches may drain basins or depressional areas, or they may collect or intercept flow from land surfaces or channelled flow from natural depressions, plow furrows, crop row furrows, and bedding systems. State drainage guides and standards and specifications have criteria regarding side slopes, grades, spacing, and depth of drainage field ditches.

Open Drain Systems

Drains should be located to fit the farm or other land use operations and should have capacity to handle the runoff and not cause harmful erosion. The drain system should cause excess water to flow readily from the land to the disposal drain. Five common drain systems are described below.

Random drain system: This type system is adapted to drainage systems on undulating land where only scattered wet areas require drainage. The ditches should be located so they intercept depressions and provide the least interference with farming operations. The ditches should be shallow and have side slopes flat enough for farm equipment to cross. Precision land forming and smoothing help to assure the removal of surface water from less permeable soil.

Parallel drain system: This type system is applicable to land where the topography is flat and regular and where uniform drainage is needed. The ditches are established parallel but not necessarily equidistant. The direction of the land slope generally determines the direction of the ditches. Field ditches are generally perpendicular to the slope, and laterals run in the direction of the slope. The location of diversions, cross slope ditches, and access roads for farming equipment can also influence the drain location. Spacing of the field ditches depends upon the water tolerance of crops, the soil hydraulic conductivity, and the uniformity of the topography. Landforming can reduce the number of ditches required by making the topography more uniform. Where possible, spacings should be adjusted to fit the number of passes of tillage and harvesting equipment.

Cross slope drain system: This system is used to drain sloping land, to prevent the accumulation of water from higher land, and to prevent the concentration of water within a field. The field ditches work best on slopes of less than 2 percent. The drain is located across the slope as straight as topography will permit. The spacing of these ditches varies with the land slope and should be based on State drainage guides. The excavated material should be placed in low areas or on the downhill side of the drain. Landforming or smoothing between the ditches improves operation of the system by preventing the concentration of flow and the occurrence of ponding.

Bedding: Bedding resembles a system of parallel field ditches with the intervening land shaped to a raised, rounded surface. This drainage system generally is used where slopes are flat and the soil is slowly permeable and where other types of drainage are not economically feasible. A bedding system generally is in small land areas and is installed using farm equipment. Beds are established to run with the land slope or in the direction of the most desirable outlet. Local information should be used to determine the width of beds, the crown height, construction method, and maintenance.

Narrow raised beds: A narrow bed system has a raised bed wide enough for single or double cropping rows to provide an aerated surface profile. This system facilitates surface water movement and aeration of the shallow root zone. When used with plastic covers for weed control, evaporation control, and nutrient management, the narrow bed system can be extremely effective for some cropping systems.

Landforming

Landforming refers to changing the land surface to ensure the orderly movement of surface water. Land smoothing and precision landforming are used in surface drainage to improve the effectiveness of the drainage system.

Smoothing

Removing small irregularities on the land surface using special equipment is termed land smoothing. It does not require a grid survey and includes operations ordinarily classed as rough grading. Land smoothing is an important practice for good surface drainage. It eliminates minor differences in field elevations and shallow depressions without changing the general contour of the ground surface appreciably. It results in better drainage, generally with fewer surface ditches, and enables farm equipment to be operated more efficiently. Smoothing also reduces ice crusting in winter.

Soils to be smoothed must have characteristics that allow small cuts. Except for isolated spots handled by prior rough grading, land smoothing operations seldom involve cuts and fills of more than 6 inches. High and low spots generally are visible to the eye without use of an engineer's level. However, sufficient spot elevations should be taken and adequate planning done to assure that water will readily move to the field ditches and laterals without ponding.

Land smoothing is accomplished best by special equipment, such as the landplane or leveller, which can work efficiently to tolerances of 0.1 foot or less. This degree of accuracy is necessary to remove irregularities from flat land. Rough grading for land smoothing can be done with farm tractors and scrapers or with a landplane. In making cuts, avoid removing all topsoil from any area. It is better to take thin layers from a larger area. Where fills from rough grading exceed 6 inches, make an allowance for settling. The allowance is related to the types of soil and soil conditions and may be given in the local technical guides. Final smoothing operations may be deferred until compaction or natural settlement of such areas has taken place.

Precision landforming

Precision landforming for drainage is reshaping the surface of the land to planned grades so that each row or plane is graded throughout its length to field ditches or other suitable outlets. The land is graded so that rows will carry surface runoff without overtopping. Row or plane grades may be varied within erosive limits to provide drainage with the least amount of grading. The length of most land smoothing equipment varies from 20 to 50 feet, with the longer length giving a more refined job. The large levellers or planes that have an overall span length of about 50 feet may require a crawler-type or 4-wheel drive tractor for power. Those that have a span length of 20 to 30 feet can be pulled with an ordinary wheel-type farm tractor. The smaller machines do a good job on any field, but require more trips over the field than the larger machine.

To facilitate smoothing operations, the ground surface should be chiselled or disked before smoothing and should be free of bulky vegetation and trash. Loosening facilitates the movement of the dirt by the leveller and mixes crop residue into the soil, thus preventing vegetation from collecting on the leveller blade. If the area is in sod, it should be tilled 3 to 6 months before smoothing operations begin, or it should be farmed in a cultivated crop for 1 year before smoothing. The roughness of the field and the number of minor depressions determine the number of passes required to produce a smooth field.

The land plane should make at least three passes; one pass along each diagonal and a final pass generally in the direction of cultivation. The first year after smoothing, settlement may occur in large depressions that were filled. On relatively flat fields, this settlement can produce pockets that collect and hold water. The field should be observed the year after smoothing. If pockets have developed, at least three more passes should be made with special emphasis on the settled areas. In areas where exposing the subsoil is necessary, you should remove and stockpile the topsoil, do the reshaping, and then replace the topsoil.

Open Drains Designing

Mains and laterals are open ditches constructed to dispose of surface and subsurface drainage water collected primarily from surface field ditches and subsurface drains. They can also intercept ground water, help to control ground water levels, or provide for leaching of saline or sodic soils. Factors affecting the size and shape of ditches are drainage runoff, hydraulic gradient, depth, bottom width, side slopes, roughness of the drain bed and banks, and limiting velocities. Local experience incorporated in the State drainage guide generally dictates the design factors for ditches.

Runoff is determined above and below the outlet of contributing ditches and streams, at points of change in the channel slope, at culverts and bridges, and at the outlet. Runoff calculations generally begin at the upper end of the drain and proceed downstream. An empirical procedure, termed the 20-40 rule, should be used in computing the required capacity for a drain below a junction with a lateral. For large drainage areas, the application of the procedure may have considerable effect on the drain design. In small areas the change in required drain capacity may be so small that the procedure need not be applied. Experience in applying the 20-40 rule will guide the designer in its use. The rules for computing the required capacity for a drain are:

Rule 1-Where the watershed area of one of the ditches is 40 to 50 percent of the total watershed area, the required capacity of the channel below the junction is determined by adding the required design capacity of each drain above the junction. This is based upon the assumption that the flows from two watersheds of about the same size may reach the junction at about the same time, and that therefore the drain capacity below the junction should be the sum of the two flows. This rule should be used in all cases for watershed areas of less than 300 acres.

Rule 2-Where the watershed area of a lateral is less than 20 percent of the total watershed area, the design capacity of the drain below the junction is determined from the drainage curve for the total watershed area.

Rule 3-Where the watershed area of a lateral is from 20 to 40 percent of the total watershed area, the discharge is proportioned from the smaller discharge at 20 percent to the larger discharge at 40 percent. In this range the discharges should be computed by both methods and the difference in cubic feet per second obtained. The design discharge for the channel below the junction should then be obtained by interpolation.

Drain Alignment

The natural topography and aesthetics should be considered in determining drain alignment. Where it is necessary to change direction of the drain or field ditch, a simple curve should be used. Curves that have a radius greater than 600 feet are desirable, but sharper curves can be used if needed to follow old ditches or swales, to decrease the waste area caused by the use of long radii curves, or to conform to ownership boundaries. Where the drain flows are small and of low velocity, gentle curvature is not as important.

Hydraulic Gradient

The hydraulic gradient is the slope of the hydraulic gradeline (water surface) and is important in determining flow velocity. Proper location of the gradeline is more important as drain flows become greater. The profile of the channel should be plotted showing the location and elevation of control points. The control points help to select the maximum elevation of the hydraulic gradeline desired for the drain. They may include, but are not limited to, the following:

- — Natural ground elevations along the route of the proposed drain.
- — Location, size, and elevation of critical low areas to be drained. These are obtained from the topographic data.
- — Hydraulic gradeline for side ditches or laterals established from the critical areas to the design drain. Plot the elevation where the side drain hydraulic gradeline meets the design drain as a control point.

— Where laterals or natural streams enter the design drain, use the same procedure as that for hydraulic gradeline for side ditches.

— Bridges across drainage ditches should not reduce the area of the design cross section. Where feasible to do so, the hydraulic gradeline should be placed 1 foot below the stringers of the bridge. The allowable head loss on culverts should be kept low. On agricultural drainage the allowable head loss generally should not exceed 0.5 foot.

— Elevations of buildings or other property within the area to be protected from overflow.

— If the drain being designed is to outlet into an existing drain or natural stream, the elevation of the water in the outlet drain or stream against which the designed drain must discharge should be used as a control point. The water surface elevation in the outlet ditch may be determined from recorded data, historic observation, or high water marks. Another method of obtaining this elevation is to determine the depth of flow in the outlet ditch by applying the same flow design basis as that used for the proposed ditch. For small outlet ditches in rather flat topography, the water elevation may be estimated at the bankfull stage.

Control points should be connected with a line on the profile. The hydraulic gradeline is drawn through or below the control points. The grades should be as long as possible and should be broken only where necessary to stay close to the control points. If the hydraulic gradeline has been well established, it will not be altered except at structures that have head losses. At these control points, the head loss will be shown upstream from the structure as a backwater curve. This will change the hydraulic gradient, although generally for only a short distance.

If the channel is in an area of flat topography and the hydraulic gradeline is located near field elevation, the hydraulic gradeline may need to be broken at culverts in amounts equal to the required head, otherwise, the backwater curve above the culvert may cause problems. The problems are compounded if several culverts are within a relatively short channel.

Different Methods for Designing

Design by reaches-Drain design may be done by reaches (specific length of ditch). One method is to design the ditch for the required capacity of the lower end of the reach and use that section throughout the reach. If this method is used, the upper end of the reach will be overdesigned. Reaches should be selected so that overdesigning is held to practical limits. To reflect the flow conditions as nearly as possible, the selection of the length of reach is important. To determine the beginning and the end of reaches, you can use one of the following:

- Tributary junctions where the required drain capacity changes.
- A break in grade of the water surface profile.
- Divided reach-An increasing drainage area may require that an otherwise long reach be divided into shorter reaches. A knowledge of where water enters the drain, as well as the amount, helps to determine how the reach should be divided.
- Bridges or culverts can be used to begin or end a reach.

After the hydraulic gradeline has been established, determine the drainage area and the required drain capacity at the upper and lower end of each reach. At this time obtain the drainage area and the required capacity for any planned structure (culvert, side inlet, grade control). This information is used in designing the structure.

Point method-Drain design by the point method is sometimes preferred. In this method the required depth of the drain is determined at the control points for the discharges at those points. This assumes that the runoff throughout the reach enters uniformly along the reach. In reality, the depth at the beginning and end of a reach differ. The depths are established at points below the hydraulic gradeline, and the bottom of the drain is drawn between the points. The slope of the drain bottom normally is not parallel to the hydraulic gradeline. At points where concentrated flow enters, a change in depth or width, or both, may be required. A transition may also be required at these points.

Drain depth

Factors that must be considered in establishing the drain depth are:

- — Depth to provide the capacity for removing the surface runoff plus freeboard.
- — Depth to provide outlet for subsurface drainage.
- — Depth to clear bridges.
- — Depth to allow for sufficient capacity after subsidence in organic soils.
- — Depth to trap sediment below the elevation of a design flow line.

Surface discharge: Sufficient depth must be provided for surface runoff to flow freely into the ditch. Where the hydraulic gradeline has been established, the depth of the ditch is measured down from the gradeline. Where the ditch bottom grade is established first, the depth is measured upward to locate the water surface. Using this last method, the water surface elevation obtained must be checked in relation to control points. The depth of each reach should be determined to meet the needs of the specific area involved.

Subsurface discharge: Ditches used to serve as outlets for subsurface drains should be deep enough to provide free outfall from the underdrain with at least 1 foot of clearance between the invert of the drain and the normal low water flow in the ditch. Ditches designed to intercept subsurface flows or to serve as an outlet for subsurface drainage should be deep enough to serve these conditions. The required depth sometimes results in the ditch capacity being in excess of the design flow requirements for surface runoff, and the actual hydraulic gradeline may be substantially lower. This often occurs at the upper end of small drainage systems.

Grade of ditch bottom

If uniform flow in the ditch is assumed, the ditch bottom grade will have the same slope as the hydraulic gradeline. The required depth of the ditch is determined and measured at points below the hydraulic gradeline. These points are then connected to find the bottom grade of the ditch. This method of locating the ditch bottom is generally satisfactory in designing a new ditch. If this method is used, you may prefer to adjust the bottom grade to eliminate the need for transitions.

In reconstruction of existing ditches, the elevation of the existing ditch bottom, the bridge footings, and the soil strata must be considered as control points in establishing the bottom of the designed ditch. At times, it may be more convenient to locate the bottom of the ditch from these control points and measure the design depth up from the bottom to locate the hydraulic gradeline (water surface). Cross sections must be used in deciding upon the hydraulic grade or the ditch bottom. Ditch bottom grades that will be erosive during normal flows are to be avoided.

Ditch side slopes

The side slopes of ditches are determined primarily by the stability of the material through which the ditch is dug and by the methods of maintenance to be practiced. Recommended sideslopes may be found in many local drainage guides. Maintenance requirements may necessitate modification. The steepest side slopes recommended for ordinary conditions for mains or laterals are as follows:

Material	Side slope (Horizontal to vertical ratio)
Solid rock, cut section	0.25:1
Loose rock or cemented gravel, cut section	0.75:1
Heavy clay, cut section	1:1
Heavy clay, fill section	2:1
Sand or silt with clay binder, cut or fill section	1.5:1
Loam	2:1
Peat, muck, and sand	1:1

Local information may indicate that steeper side slopes can be used in certain soils; however, flatter side slopes may be desirable for more satisfactory and economical maintenance.

Velocity in ditches

The velocity in a ditch is ideal if neither scouring nor sedimentation occur. Because flows in most ditches are intermittent, the velocity will fluctuate. Ditches are generally designed for the maximum design flow and the allowable average velocity in the ditch section. If the velocity is above these limits, scouring and erosion may take place. Because raw, newly dug ditches may have a lower roughness coefficient (Manning's n) than they will have after some aging of the ditch takes place, they may have higher than design velocities.

Where vegetation on the ditch side slopes is slow in developing, the limiting velocities may need to be reduced. Minimum design velocities should not be less than 1.4 feet per second to prevent sedimentation. Short field ditches may be used to capture sediment for a water quality benefit. In this case, design velocity should be less than 1.4 foot per second with provision to return sediment to the field.

Manning's equation is used in determining the average velocity in a ditch section.

$$V = \frac{1.486}{n} R^{0.667} S^{0.5}$$

where:

V = velocity (ft/s)

n = roughness coefficient

R = hydraulic radius (ft) = A/P

S = slope (ft/ft)

A = cross-sectional area below hydraulic gradeline (ft^2)

P = wetted perimeter (ft)

The value of n is a factor in Manning's formula for computing velocity. It indicates not only the roughness of the sides and bottom of the channel, but also other types of irregularities of the channel, such as alignment and vegetation. The value of n is used to indicate the net effect of all factors causing retardation of flow. The selection requires judgement in evaluating the material in which the channel is constructed, the irregularity of surfaces of the ditch sides and bottom, the variations in the shape and size of cross sections, and the obstructions, vegetation, and meandering of the ditch.

The machinery used for construction of the ditch should be considered in the selection of ditch bottom width. A bulldozer or blade equipment is used to construct V-shaped ditches. Flat bottom ditches frequently are designed if scrapers, hydraulic hoes, or draglines are to be used to construct the ditch. Depth of ditch and soil conditions affect the type of equipment used. Specified minimum bottom widths are often based on the available equipment.

Depth and bottom width

The most economical ditch cross-section approaches that of a semicircle. A deep, narrow ditch generally carries more water than a wide, shallow ditch of the same cross-sectional area. An excessively wide, shallow ditch tends to develop sand or silt bars, which cause ditch meandering and bank cutting, and a fairly deep, narrow ditch tends to increase velocities and reduce siltation and meandering. Because the crosssection selected is a matter of judgment, all factors involved should be considered. Ditches shall be designed to be stable. In some cases economy and hydraulic efficiency must be sacrificed in the interest of ditch stability and maintenance.

The volume (Q) of water passing a ditch cross section is calculated in cubic feet per second (ft^3/s) and is the product of the flow area cross section (A) in square feet (ft2) and the average velocity in the cross section (V) expressed in feet per second (ft/s). The formula is:

$$Q = AV$$

Various curves, tables, and computer software, all based on Manning's formula for velocity, have been prepared to determine ditch capacities.

Ditch berms

Berms should be designed to:

— provide roadways for maintenance equipment;

— provide work areas;

— facilitate spoil-bank spreading;

— prevent excavated material from washing back into ditches; and

— prevent sloughing of ditchbanks caused by placing heavy loads too near the edge of the ditch.

The recommended minimum berm widths are as follows:

Ditch depth	Minimum berm width
ft	ft
2 - 6	8
6 - 8	10
>8	15

Spoil banks

Spoil-bank levelling or shaping is a common and desirable practice. The degree of levelling, the placing of the spoil, and other practices related to the spoil generally are determined locally and are specified in local drainage guides or State standards and specifications.

Bridges and culverts

Design criteria for structures required for drainage ditches, or where irrigation canals cross drainage ditches, are specified by the authority responsible for the structure. The capacity requirement for the structure may be for flood flows that are much more intense than the drainage requirement. On some township, private, and field roads the only requirement may be that the structure carry drainage flow. The following information is limited to this type of structure. The structure must meet two requirements:

— They must be of sufficient size and located so as to pass the design flow within the allowable head loss.

— They must have adequate strength, size, and durability to meet the requirements of traffic.

The following formula can be used to compute the minimum culvert length without headwalls.

$$L = W + 2SH$$

where:

L = minimum culvert length

W = top width of fill over culvert, not less than levee top width

S = side slopes of fill over culvert

H = height of fill measured from culvert invert

Existing structures should be measured to determine their capacity. An existing structure may be considered adequate if it will pass the design drainage flow with a head that does not cause overbank flow above the structure. As a safety factor, new culvert installations generally are designed for 25 percent more capacity than the ditch design. A new bridge should be designed to span the ditch, have the bottom of the stringers placed at least 1 foot above the hydraulic gradeline, and preferably have no piers placed in the centre of the ditch. The hydraulic gradeline at structures does not need to be broken as long as the head loss required to pass the design flow does not cause overbank flooding above the structure.

The hydraulic gradeline should be adjusted if such flooding will occur. The gradeline may be broken and dropped down at the structure at an amount equal to the head loss. Head loss is negligible through bridges where the channel cross section is not restricted. Head loss through culverts ordinarily should not exceed 0.5 foot. It can be reduced by increasing the size of the structure.

Culvert flow

Culverts are used for several types of flow. Detailed knowledge of hydraulics is necessary under some situations for the design of culverts. The following situations commonly occur where culverts are used:

- Culverts flowing full with both ends submerged.
- Culverts flowing full with unsubmerged or free discharge.
- Culvert flow limited by culvert entrance conditions.

Ditch junctions

The bottom grades of ditches having about the same depth and capacity should be designed to meet at or near the same elevation. The bottom of a shallow, small capacity ditch may be designed to meet a larger ditch at or near the normal or low flow elevation of the larger ditch. A transition is designed where a shallow ditch enters a much deeper ditch. Before beginning a transition, the grade of the shallow ditch generally is designed 10 to 100 feet upstream on a zero grade at the elevation of the deeper ditch. The transition should be on a nonerosive grade not to exceed 1 percent. Where the difference in the elevation of the ditch gradelines is considerable and transition grades seem impractical, a structure should be used to control the drop from the shallow ditch to the deeper ditch.

Surface water inlets to ditches

All drainage into mains and laterals should be through planned inlets, rather than at random, which can cause rills or severe bank erosion. This may be accomplished by installing chutes, drop spillways, pipe drop inlet spillways, culverts, or other suitable structures.

Swinging watergates, cattle guards, and ramps

Where applicable, watergates, cattle guards, and ramps should be used on open ditches to manage livestock and to protect the ditches.

Construction of Drainage

Open Drain Layout

The amount of staking required depends upon the topography, the size of ditch, the type of equipment used, and the experience of the contractor or cooperator. A centerline stake, slope stakes, and offset reference stakes may be set at every station. Sometimes the centerline cut or fill is marked on the centerline stake. In many cases staking is not required at every station, but a sufficient number of stakes should be set to obtain the intended result. The method used should conform to state requirements.

All culverts and bridges should be installed or rehabilitated immediately after the site has been excavated. The structures should be installed as planned. The bottom of ditches should be rounded to conform to the shape of the culverts. Where multiple pipe culverts are used, the space between the barrels should be at least half the diameter of the culvert. All backfill should be carefully and firmly tamped.

Swinging Watergates

Swinging watergates should be constructed of light, durable material. They should be hung so that they do not swing through too great an arc before the bottom of the gate rises to the elevation of the water surface. To prevent the gate from becoming grass bound, a clearance of about 6 inches between the gate and the bottom and sides of the ditch is needed.

Surface Water Inlets

Pipe overfall structures generally discharge into areas recessed in the banks of the outlet ditch. This is especially necessary if the outlet is a flowing stream. If they are installed in this manner, they will not be damaged by the movement of water, ice, and debris in the outlet; and the flow in the outlet will not be retarded. The installation should be completed by adequately tamping the soil around the pipe. To prevent a failure by washout, the fill over the pipe should be brought up high enough and along the pipe far enough to force any possible overflow water to the sides. This is generally referred to as an island method of installation.

The installation of inlet structures on fills should be avoided. The pipe needs to be well bedded-the bottom part of the excavation should conform closely to the shape of the pipe. This fine grading should extend up the sides of the pipe to a point where the backfill can be easily reached with a hand or mechanical tamper. All joints should be watertight.

Berms, Spoil Banks, and Seeding

The use of the practices of leaving berms on ditches, levelling of spoil, and seeding ditch slopes, berms, and levelled areas varies with the locality. The spoil should be shaped to facilitate maintenance. Where no berm is left and spoil is shaped into a road, the height of the road should be limited to that which will not cause sloughing of the ditchbanks. In all cases, the spoil should be shaped so that the minimum amount of water flows directly back into the ditch. Consideration should be given to controlling erosion of the spoil. State standards for the above items should be used.

Maintenance Programmes

A definite maintenance programme should be agreed upon when the drainage system is planned because the maintenance methods to be used can materially affect the design. A good maintenance plan should include the practices to be used as well as the approximate time of the year when specific practices are applicable. Unless the growth of vegetation and silting are controlled by regular maintenance, they may quickly decrease the effectiveness of the drain. The capacity of an open drain can be reduced by as much as 50 percent in 1 year by sediment and a heavy growth of weeds or brush.

Mowing

Mowing to control vegetation on spoil banks, berms, and drains is economical and effective if the side slopes are not too steep. For safety, side slopes on which farm tractors are to be operated should be made 3:1 or flatter-4:1 slopes are preferred. Side slopes as steep as 1:1 can be mowed using mowers mounted on hydraulically operated booms. Timing and frequency of mowing should consider wildlife values, weed control, and considerations for controlling other vegetative conditions, such as to avoid snow entrapment along traffic lanes and volunteer woody growth.

Burning

Burning in winter or early spring, when the vegetation is dry and the ground is wet, is sometimes used to control undesirable vegetation in drainage ditches. Bridges, plastic pipe outlets, fences, and other property must be protected from burning. Any burning activity should be in compliance with State and local laws. Impacts on air quality should be a major consideration.

Chemicals

Chemicals to control undesirable vegetative growth have produced some excellent results and are used by land users and drainage enterprises. Caution should be used in their application to prevent impact to wildlife, crop damage, and water pollution from the drifting chemicals. Broadleaf crops and some truck crops are particularly susceptible to damage. Information on appropriate chemicals is available from local dealers or current USDA publications.

Major chemical companies have prepared considerable information relative to usage of specific products. Because of environmental dangers associated with the use of herbicides or burning, the search for other solutions is ongoing. In open channels with permanent water, fish known to consume large quantities of vegetation, such as the Tilapia, are used with good results. This application is very limited; however, a greater awareness of this alternative and a continuing search for acceptable methods are needed.

Land preparation practices require year-to-year maintenance to retain their efficiency. Once the field is properly prepared to achieve good surface drainage, the ordinary crop cycle involving tillage, planting, cultivating, and harvesting, along with the wind and water action during the year, can disturb the surface enough to impound water and cause crop damage.

Because cropping cycle irregularities, such as implement scars, need to be erased before each crop is planted, a leveller or plane should be operated over the area each year. This operation also will take care of settlement in the fill areas and provide a base for a good seedbed.

3

Subsurface Drainage

Subsurface drainage removes or controls free water from the soil surface and below the surface of the ground. The principal function of subsurface drainage is to prevent, eliminate, or control a high water table. Lowering the water table can improve growing conditions for crops, the condition of the soil surface, and trafficability on the field as well as around the farmstead. It also facilitates tillage practices.

Subsurface drainage also provides a period to apply agricultural wastes during high and prolonged rainfall seasons, and it maintains water intake and soil profile storage capacity, thereby reducing runoff during high rainfall periods. This type of drainage functions in irrigated areas to control saline and sodic soil conditions by removing excess salt accumulations, to provide for subirrigation, to help control seepage from canals and laterals, and to remove excess irrigation water from sources upslope as well as onsite. Subsurface drainage is accomplished by various kinds of buried or open drains.

Subsurface drainage is applicable to saturated soil conditions where it is physically and economically feasible to use buried conduits to remove or control free water from the root zone. The need for and the design of subsurface drainage systems are related to the amount of excess water entering the soil from rainfall, irrigation, or canal seepage; the permeability of the soil and underlying subsoil material; and the crop requirements. In soils with slow permeability that causes water to flow slowly into the drain, the drains must be closely spaced.

Consequently, installation may be considered too expensive for use of subsurface drains. Soils must have sufficient depth and permeability to permit installation of an effective and economical subsurface drainage system. Some sandy soils and peat and muck have large pore spaces that allow rapid movement of water. Wetness occurs in these soils because of a high water table, particularly in the spring in nonirrigated areas, late in summer, or during the irrigation period. For maximum crop yields, the wetness problem must be corrected by drainage. These soils can be successfully drained. Some fine sand soils have insufficient colloidal material to hold the sand particles together. This can cause excessive movement of the particles into the drains. Special precautions, such as filters or envelopes, are often required.

In highly permeable, coarse sands and some peat soils, excessive lowering of the water table causes a moisture deficiency during periods of drought. Such soils have limited capillary rise and are unable to deliver water up into the plant root zone of certain crops if the water table falls much below the root zone. Water table control systems should be used for these conditions. Other soil conditions make construction of drains hazardous or impractical. In some soils, boulders or stones make drainage costs prohibitive. In others, the topsoil is satisfactory, but it is underlain by unstable sand at the depth where drains should be installed, thus making installation more difficult.

A chemical action, which takes place in soils that have glauconite, iron oxide, or magnesium oxide, can cause drain joints or perforations to seal over. In soils where iron is present in soluble ferrous form, ochre deposits in drain lines can be a serious problem. If the problem is recognised, it can be solved by making adjustments in design and maintenance of the drainage system. Ochre is formed as a combination of bacterial slimes, organic material, and oxidised iron. It is a highly visible, red, gelatinous, iron sludge that often occurs in the valleys of the corrugations of drain tubing as well as at the drain outlets.

Soil conditions that contribute to ochre formation have been identified throughout the United States. Studies have shown a relationship among soil types, iron ochre, and related sludge deposits in subsurface drain lines. Four known sludge deposits are associated with bacterial activity in subsurface drains-ochre, manganese deposits, sulphur slime, and iron sulphide. Iron deposits, collectively named ochre, are the most serious and widespread of the sludge deposits.

The ochre and associated slimes are a sticky mass that is generally red, yellow, or tan. This sticky mass can clog drain entry slots, drain envelopes, and the valleys of the corrugations between envelope and inlet slots. Such elements as aluminum, magnesium, sulphur, and silicon are often present. The soils that tend to show the most hazard for ochre formation are fine sand, silty sand, muck or peat that has organic pans (spodic horizons), and mineral soils that have mixed organic matter. Those that have the least potential hazard of ochre are silty clay and clay loam.

Sites used for spray irrigation of sewage effluent and cannery plant wastes generally furnish sufficient iron and energy for reduction reactions; therefore, the potential hazard of ochre deposits is serious. In certain areas of the western United States, manganese, when present under suitable conditions in the ground water, can form a drain-clogging, gelatinous black deposit. Manganese has not been a serious problem in the Eastern United States. Sulphur slime is a yellow to white stringy deposit formed by the oxidation of the hydrogen sulphide in ground water. This slime has not been a serious problem in most agricultural drains. It is most frequent in muck soils. It may be on sites designed for subirrigation if the well water used for irrigation contains hydrogen sulphide.

Applications

Field Drainage

Relief drains are those installed to remove excess ground water percolating through the soil or to control a high water table. They should systematically lower the water table for an area. The drains may be aligned parallel or perpendicular to the direction of ground water flow. Relief drains will develop similar drawdown conditions on either side of the drain, and, if the soil is homogeneous, the water table on either side will be the same at equal distances from the drain.

Interception Drainage

Interception drains are installed at right angles to the flow of ground water to intercept subsurface flows. The drainage is applicable to broad, flat areas that are wet because of seepage from canals or adjoining highlands. Drains for interception of seep planes must be located properly to dry wet areas caused by upslope water. Seep planes are first located by soil borings, and then the drain is located so that continuous interception of such seep planes, adequate soil cover over the drain, and uniform grade to an available outlet are established. In steeply graded depressions or draws, a layout may include a main or submain drain in the draw or to one side of the draw, with the interceptor lines across the slope on grades slightly off contour.

Drainage by Pumping

The objective of all subsurface drainage is to remove or control excess water from the root zone of the crop. This generally is accomplished by installing subsurface drains or open ditches. Water table levels may also be controlled by pumping from the ground water reservoir to maintain the upper limit of the water table level. In some irrigated areas where irrigation water is obtained from wells, irrigation and drainage may both be affected by the pumping of wells. This combination of practices is limited to areas where the soil has low salinity and a proper salt balance can be maintained.

In salty areas where drainage is accomplished by pumping, the drain water is generally discharged into a natural outlet or planned disposal system and not directly reused for irrigation. In some cases the drain effluent can be used conjunctively with a water source of higher quality, making the water suitable for irrigation use. Planning pump drainage can be quite complex. Detailed information on the geologic conditions and the permeability of soil and subsoil material is important. Design involves anticipating what the shape and configuration of the cone of depression will be after pumping. This, in turn, involves spacing of wells with respect to their areas of influence to obtain the desired drawdown over the area to be drained. Experience with this type of drainage indicates that it is costly and consideration of its use may be limited to high-producing lands that have a high net return per acre.

Mole drainage

Mole drains are cylindrical channels artificially produced in the subsoil without trenching from the surface. Various kinds of plastic liners and methods of installation of the liners have been tried with varying degrees of success. The object of the lining is to extend the life of the mole channel. Mole drainage is used in organic soils and soils with a dense, impervious, fine-textured subsoil and normally in more undulating areas. The problem is not the control of a ground water table (which may be very deep), but the removal of excess water from the field surface or from the topsoil.

The water reaches the mole channel mainly through fissures and cracks formed during installation of the mole opening. The outflow from mole drainage systems differs considerably from that of subsurface drainage systems controlling the ground water table. Normally, the outflow from mole drainage systems show a quick response to rainfall or irrigation water; thus, as the water application or rainfall event ceases, the outflow ends quickly. The time lag between water intake at the surface and drain outflow is a few hours at most. Experience with use of mole drainage in the United States is so variable that it is impractical to describe its use here.

Water Table Management

Water table management is the operation and management of a ground water table to maintain proper soil moisture for optimum plant growth, to sustain or improve water quality, and to conserve water. It is the operation of a subsurface drainage system for the purpose of lowering the water table below the root zone during wet periods (drainage), raising the water table during dry periods (subirrigation), and maintaining the water table during transition (controlled drainage).

Drainage for Salinity Control

Salinity control is practiced for both irrigated cropland and dryland farming operations. The same salt balance principles apply, but the source of the salts and the water source and management must be understood and treated appropriately. Drainage for salinity control on dry land commonly intercepts and removes excess ground water upslope of areas in which the subsoil has a high salt content. The excess water may come during high precipitation periods, noncrop growing periods, or during heavy snow accumulation. Removal of the excess water above the seep prevents the accumulation of salts left by the surfacing of the fluctuating water table. Drainage facilities may be used alone or in combination with special agronomic practices to manage the excess water.

Dryland salinity management: Interception drainage may play an important role in management of ground water. Open interceptor drains may be effective and better suit a particular site condition than a subsurface drain. Saline or sodic conditions can also occur on nonirrigated land at or near sea level in coastal areas. Controlled drainage structures or pumping systems may be needed to make a drainage system function satisfactorily.

Salinity management for irrigated conditions: The management of saline conditions in irrigated agriculture is practiced more extensively and perhaps better understood. The source of the salts is generally from the irrigation water. As the water is used by evapotranspiration, the salts are either leached below or out of the root zone, or they are left to accumulate in the root zone. If the salts are left to accumulate in the soil profile, a noticeable reduction in yield occurs, depending on the specific crop tolerance.

A major application of subsurface drainage in nonagricultural settings is for foundation, basement, and lawn areas of residential homes. Saline and sodic problems are most common in arid and semi-arid climates. Where salt crusts are visible or salinity is suspected, the soil should be sampled at various depths. A record of the species and condition of plant cover is important. Soil investigations should be conducted to estimate the maximum potential for ochre development in the planned drainage system.

Biological and Mineral Clogging

The ferrous iron content of the ground water flowing into a drain is a reliable indicator of the potential for ochre development. Soluble ferrous iron flowing in ground water enters a different environment as it approaches the drain and passes through the drain envelope. If the level of oxygen is low, certain filamentous and rod-shaped bacteria can precipitate insoluble ferric iron and cause its incorporation into the complex called ochre. The amounts of iron in ground water that can stimulate bacteria to produce ochre can be as low as 0.2 ppm.

Laboratory and field methods are available to estimate the ochre potential for a given site. Of particular importance is whether ochre may be permanent or temporary. Temporary ochre occurs rapidly, usually during the first few months after drain installation. If the drains can be cleaned or maintained in functional order, the ochre problem may gradually disappear as the content of iron flowing to the drains is reduced. Such soil environments must be low in residual organic energy sources to prevent the continual release of iron during short-term flooding.

Permanent ochre problems have been found in profiles with extensive residual iron, such as cemented iron subhorizons or rocks, and from iron flowing in from surrounding areas. Many factors influence ochre deposition, including the pH, type, temperature, and reducing conditions of the soil. Certain onsite observations may give clues to potential ochre formation before a drainage system is installed.

Surface water in canals may contain an oil-like film that is iron and may contain Leptothrix bacterial filaments. Gelatinous ochre may form on the ditchbanks or bottom of canals. Ochre may also form within layers of the soil. Iron concretions, sometimes called iron rocks, are in some areas. The presence of spodic horizons (organic layers) suggests ochre potential; and most organic soils, such as mucks, have some potential for ochre problems. If a site has potential for ochre deposits, certain planning and design practices should be followed to minimise this hazard to the system. No economical, longterm method for effectively controlling this problem is known.

For sandy soils where a filter is necessary, a graded gravel envelope is best, although it can become clogged under conditions of severe ochre potential. When synthetic fabrics were evaluated for ochre clogging, the knitted polyester material showed the least clogging. A submerged outlet may be successfully used to minimise ochre development with the entire drain permanently under water. The line should be completely under water over its entire length throughout the year. This may require that the drains be on flat grade. The depth of ground water over the drain should be at least 1 foot.

Herringbone or similar drain designs should have entry ports for jet cleaning. Use drain pipe that has the largest slots or holes allowed within the limits of drain pipe and envelope standards. Slots or holes should be cleanly cut and should not have fragments of plastic to which ochre can adhere. Both smooth bore and corrugated pipes can accumulate ochre.

Water table contour maps and depth to water table maps are very useful in planning subsurface drainage. The elevation of the water table at selected points, as obtained from borings, is plotted on a topographic map of the area. By interpolation, lines of equal water table are drawn. The result is a contour map of the water table. Where a ground surface contour crosses a water table contour, the depth to the ground water is the difference in elevation of the two contours. Areas that have a range of depths can be delineated on the map. The direction of ground water flow and extent of high water table areas are also shown on the map.

Outlets

System outlets: The starting point in planning a subsurface drainage system is normally the location of the outlet. Drains may discharge by gravity into natural or constructed drains. Any of these outlets are suitable if they are deep enough and of sufficient capacity to carry all the drainage water from the entire drainage system. The adequacy of the outlet should be determined before proceeding with the design of the system.

Capacity and depth of open ditch outlets: The outlet ditch must have the capacity to remove the drainage runoff from its watershed quickly enough to prevent crop damage. It should be deep enough to allow at least 1 foot of clearance between the flow line of the drain and the normal low water stage in the ditch when drains are installed at the specified depth.

Capacity and depth of subsurface outlets: If existing subsurface drains are used for the outlet, they should be in good condition and working properly. The main drain should have sufficient capacity to handle the proposed drainage system in addition to other systems it serves, and it should be deep enough to permit the new system to be installed at the depth specified.

Pump outlet: An outlet by pumping should be considered for drainage sites where a gravity outlet is not available.

Vertical drain outlet: A vertical drain is a well, pipe, pit, or bore, drilled into porous underlying strata, into which drainage water can be discharged. It is sometimes called a drainage well.

Wells tap permanent sources of ground water for livestock and domestic use. In some areas, shallow wells are the only reasonable and available water source for domestic use and must be protected from contaminants in agricultural drainage water. Public health laws in some States regulate the use of wells for drainage, and in many cases, prohibit this use because of the potential contamination of ground water and the danger to public health. However, in some parts of the country vertical drainage is a satisfactory solution to drainage water disposal and deep ground water recharge. Where the possibility of using wells for a drainage outlet is considered, an engineer or geologist, or both, should assist in the investigation and planning.

Designing of a Field Drainage System

To plan a field drainage system, a pattern should be

selected that fits the topography, sources of excess water, and other field conditions. The following basic systems may be considered.

Random System

A random field drainage system is used where the topography is undulating or rolling and has isolated wet areas. The main drain is generally placed in the lowest natural depression, and smaller drains branch off to tap the wet areas. Because such drains often become outlets for a more complete system established in the higher areas of the field, the depth, location, and capacity of the random lines should be considered as part of a complete drainage system. Generally, the logical location of these drains obviously fit the topography.

Parallel System

The parallel field drainage system consists of laterals that are perpendicular to the main drain. Variations of this system are often used with other patterns. In many cases, the parallel system is desirable because it provides intensive drainage of a given field or area. It can also be used in depressional or low areas that can be graded before installation of the system.

Herringbone System

The herringbone field drainage system consists of laterals that enter the main drain at an angle, generally from both sides. If site conditions permit, this system can be used in place of the parallel system. It can also be used where the main is located on the major slope and the lateral grade is obtained by angling the laterals upslope. This pattern may be used with other patterns in laying out a composite system in small or irregular areas.

Drainage Coefficient

Humid areas: Drains should have sufficient capacity to remove excess water from minor surface depressions and the major part of the root zone within 24 to 48 hours after rainfall ceases. The required amount of water to be removed in some specified time is the drainage coefficient. For field drainage, it is expressed as inches of water depth to be removed over a safe period of time, generally 24 hours, or as an inflow rate per unit length of drain. Because of the differences in soil permeability, climate, and crops, as well as the manner in which water may enter the drain (i.e., all from subsurface flow or part from subsurface flow and part from surface inlets), the coefficient must be modified to fit site conditions in accordance with local drainage guides or within the approximate limits as follows:

- Where drainage is uniform over an area through a systematic pattern of drains and surface water is removed by field ditches or watercourses, the coefficient should be within the range as shown in table 1. Figure only the area to be drained as the drainage area.
- Where surface water, including roof runoff, must be admitted through surface inlets to the drain, an adjustment in the capacity of the drain is required. Runoff from an area served by a surface water inlet takes place soon after the rainfall and enters the drain ahead of the ground water. In short drainage lines or small systems that have only one or two inlets, the size of the drain may not need to be increased. As systems become larger or the inlets more numerous, an adjustment to the drainage coefficient should be made. The timing of the surface water flow in relation to the entrance of

ground water into the drain should be the basis for increasing the coefficient over those shown in table 1.

Table 1. Drainage coefficients

Soil	Field crops	Truck crops
	(inches to be removed in 24 hours)	
Mineral	3/8 – 1/2	1/2 – 3/4
Organic	1/2 – 3/4	3/4 – 1 1/2

— A higher coefficient than those given in table 1 is sometimes necessary to hold crop damage to a minimum.

Arid areas: In areas where rainfall is light, drainage coefficients applicable to local areas are variable and depend upon the quality and amount of irrigation water applied, methods of irrigation, crops to be grown, and characteristics of the soil. Local drainage guides and other information should be used to determine drainage coefficients. However, where experience is lacking, the following formula can be used to estimate drainage coefficients. The deep percolation volume must be accounted for, including leach water that must be added to maintain a favourable salt balance in the soil and the water that is a result of inefficient or excess irrigation. This formula does not account for upslope water sources or upward flux from ground water.

$$q = \frac{\frac{P+C}{100}(I)}{24F}$$

where:

q = drainage coefficient, inches/hour

P = deep percolation from irrigation including leaching requirement, percent (based on consumptive use studies)

C = field canal losses, percent

i = irrigation application, inches

F = frequency of application, days

Hydraulic Conductivity

Any drainage survey should involve a hydraulic conductivity (K) investigation in the field. Twice the depth of drains or the top 7 feet of the soil profile, whichever is less, can be surveyed using the auger hole method. When using this method, the observation points should be selected according to available soil maps and information on soil morphology and its lateral variation pattern in relation to physiography. Typically, the intensity of a hydraulic conductivity survey for the purpose of project design varies from one hole per 10 acres to one hole per 20 acres, depending on the degree of homogeneity in the obtainable values.

Maps should be prepared at a suitable scale showing the K values differentiated, if possible, for hydraulic conductivity above and below drain level and also showing depths to an impermeable stratum if one exists. All auger holes should be drilled at least to the depth the drains will be installed. It is advisable to extend one hole in 10 to the relatively impermeable layer (barrier), three holes in 10 to a depth below the drain level, and six holes in 10 to the depth of the drain level.

A farm or large field on which the values of hydraulic conductivity are quite uniform may be given an average value by simple calculation of the arithmetic mean. Sometimes the farm or field must be divided into sections, each having values that are uniform but whose averages are different. Areas with an average hydraulic conductivity of 0.3 foot per day or less normally are not provided with subsurface drainage. The chosen figure, 0.3 foot per day, is based on economic considerations. Hydraulic conductivity is subject to change with time as a result of alternate drying and wetting of the soil or by changes in the soil structure through chemical or mechanical means, as well as the farming practices.

Isolated spots of very low hydraulic conductivity may be encountered on farms that are otherwise reasonably permeable. These spots should not be included in the average. The drainage of such localised impermeable mounds is best achieved by draining the surrounding, more permeable soil into which these mounds can then dewater. Installation of subsurface drainage systems in such impermeable soil should be avoided. In some situations shallow, open field ditches may be feasible. The landowner decides on the feasibility of installing a drainage system with an appropriate drain spacing. Some over or under drainage is inevitable, thus the decisionmaker should be informed of these potential irregularities as decisions are made on spacings.

Impermeable Layer

The impermeable layer (barrier) is not necessarily a more fine textured subsoil. It may be a product of other factors, such as consolidation by the weight of glaciers or a stratification of alluvial sediment. The impermeable layer can be defined as a layer with a hydraulic conductivity value of one-tenth or less than that of the soil stratum containing the water table through which the drainage water moves towards the field drains. In that regard, the effects of possible up or downward seepage through the impermeable layer have been neglected, and any lateral seepage (resulting from canal leakage or hillside seepage) is not considered. The influence of an impermeable layer on the behaviour of a ground water table depends on its depth below the level of field drains and on the drain spacing.

The flow pattern of the water moving toward the drain is altered drastically by the impermeable layer if the drain level above the impermeable layer is less than a fourth the spacing between drains. The drains need to be placed closer together to achieve the effect they would have in a deep permeable soil. However, if the depth of the impermeable layer below drain level exceeds a fourth of the drain spacing, the flow system can be treated as if such a layer was absent.

Depth and Spacing

The two basic formulas to determine spacing are the steady state and the nonsteady state, also called transient state. A basic form of the steady state equation, referred to as the ellipse equation, is presented in this section. The importance of a good drainage formula for depth and spacing is often overestimated. The transient state formula may better relate actual conditions than the ellipse equation, but the accuracy of results from any of the accepted equations is controlled primarily by the accuracy of the assumed or measured parameters used in the calculations.

The accuracy of the input parameters, especially hydraulic conductivity, may be in error by as much as 20 percent; therefore, a minor difference in the result of one equation over another is not that important. Sensitivity analysis of input parameters has been done and documented in conjunction with evaluations of

DRAINMOD. For these reasons and to simplify use, it is justifiable to use the presented steady state equation. The DRAINMOD computer program can satisfy the need for a more precise analysis of the water table fluctuations and resulting crop impact analysis.

Generally, the greater the depth of a field drain, the wider the spacing can be between drains. Other factors that help determine the depth and spacing are the soil, climate, subsurface barriers, frost depth, and crop requirements. Where the experience with drain installations is extensive, the depth and spacing of drainage systems in various soils are already well established. Drains should be deep enough to provide protection against tillage operations, equipment loading, and frost. Initial settlement in organic soils should be considered in depth selection.

Main and submain drains must be deep enough to provide the specified depth for outlets of lateral drains. Also, the maximum depth at which drains can be laid to withstand trench loading varies with the width of the trench and the crushing strength of the drain to be used. In areas where drainage installations and knowledge of effective spacings are limited, the ellipse equation can be used to determine drain spacing.

$$S = \left[\frac{4k(b^2 - a^2}{q}\right]^{0.5}$$

where:

S = spacing of drains, feet,

K = hydraulic conductivity, inches/hour

b = distance from the drawdown curve to barrier stratum at midpoint between the drains, feet

a = distance from drains to the barrier, feet

q = drainage coefficient, inches/hour

To control salinity, drains generally are placed deeper in arid and semiarid irrigated areas than in humid areas. The minimum effective depth is 6 feet for medium textured soils and up to 7 feet for fine textured soils, assuming there is no barrier above this depth. This is necessary to maintain the ground water at a depth of 4 to 5 feet midway between the drains. In most cases, the depth of the laterals depends upon the outlet depth, slope of the ground surface, and depth to any barrier.

The depth capacity of the drain-laying machine may limit the depth of the drain unless provisions are made for ramping part of the line. In ramping, a depressed roadway is excavated for the machine to obtain the needed depth in the deeper reaches of the drain line. If the outlet for the drain is not deep enough to permit its installation to the required depth, then outletting into a sump with an automatic pump cycling may be considered.

In most cases subsurface drains should not be placed under an impermeable layer, and they should be located within the most permeable layer. Economic depth should also receive consideration. Economic depth figures can easily be developed locally by evaluating changes in the cost for different depth and spacing ratios for each major soil type. The maximum depth at which subsurface drains can be safely laid varies with the crushing strength of the drain, type of bedding or foundation on which the drain will be laid, the width of the trench, and the weight of the backfill. Temperature is also a factor to consider for installation and loading of thermoplastic pipe.

In the ordinary bedding method, the trench bottom is shaped by the shoe of the trenching machine to provide a reasonably close fit for a width of about 50 percent of the conduit diameter. The remainder of the conduit is covered with granular material to a height of at least 0.5 feet above its top or topsoil is placed to fill all spaces under and around the drain. The present method of installing drains using trenching machines is effective, and for practical purposes ordinary bedding may be assumed. The bedding (ordinary) allows the drain to be installed in a deeper or wider trench than if no attention had been paid to the bedding of the drain.

For computation of maximum allowable loads on subsurface drains, use the trench and bedding conditions specified and the crushing strength of the kind and class of drain. The design load on the conduit should be based on a combination of equipment loads and trench loads. Equipment loads are based on the maximum expected wheel loads for the equipment to be used, the minimum height of cover over the conduit, and the trench width. Equipment loads on the conduit may be neglected if the depth of cover exceeds 6 feet. Trench loads are based on the type of backfill over the conduit, the width of the trench, and the unit weight of the backfill material.

The size of drains depends upon the required flow and the grade on which they are laid. The required flow is determined from the drainage coefficient and the area or length of drains contributing flow, plus any allowances for concentrated flow entering from the surface, springs, or other sources. The contributing drainage area for a complete drainage system is about the same as the total length of all contributing lines multiplied by the spacing between such lines. Random drains in poorly drained depressions are often used later as main drains for a more complete drainage system. Where such expansion is likely, the additional area that such drains would serve should be included in determining the size of the initial random line. Where surface water is admitted directly into a drain by surface inlets, the entire watershed contributing to the inlet should be included. Flow from such watersheds often can be reduced by diversion ditches.

4

Water Table Management

While there are instances where ecosystems have been damaged by poor quality drainage water, it is also possible to use subsurface drainage systems beneficially to improve water quality. By managing the water table, through controlled drainage or sub-irrigation, nitrate concentrations in drainage effluent can be significantly reduced. Downstream flood flows can also be reduced. Water table management could therefore be viewed as a best management practice. Skaggs and Gilliam and Madramootoo et al. have shown that sub-irrigation and controlled drainage not only increase crop yields, but also enhance denitrification.

Water table management is best suited to flat lands, and can be achieved by a simple modification to the outlet of subsurface drainage systems. It is important to recognise that while technologies may be available to minimise water quality impacts from drainage effluent, institutional mechanisms must also be put in place to ensure that the technologies can be implemented. Policies and programmes, including legal and monitoring aspects, require an institutional framework.

Water table management systems are currently being tested in some parts of Europe and Asia. The practice is yet to be evaluated in the arid and semi-arid regions, although there is potential for water table management in these regions. However, proper management of the water table is required to ensure that re-salinisation of the crop root zone does not occur. Early attempts at sub-irrigation in arid and semi-arid regions failed because of soil salinisation.

In the humid regions of North America, water table management has evolved on lands which were first improved with surface and subsurface drainage. Drainage improvements were initially required on these poorly drained soils to increase crop productivity and provide better soil conditions for field machine trafficability. With concerns about the quality and impacts of agricultural drainage water on marine ecosystems, as well as nitrate contamination of drinking water, the drainage systems were modified to reduce the concentration of pollutants in the drainage water. Water table management has been practised for several years on organic and sandy soils in humid regions to reduce drought stress of high value crops.

There are two forms of water table management: controlled drainage and sub-irrigation. Controlled drainage restricts the discharge from a subsurface drain outlet or open main drain, resulting in a higher field water table. The water table drops naturally over time due to evaporation and deep seepage, and is only raised if there is precipitation. With sub-irrigation, water is pumped slowly and nearly continually into open ditches or a subsurface drainage system to maintain a near constant water table. When large rainfalls occur and the field water table rises above the desired level, the irrigation pump is stopped. The excess water then drains from a control structure in the ditch or drain outlet. Such systems are shown in Figure 1.

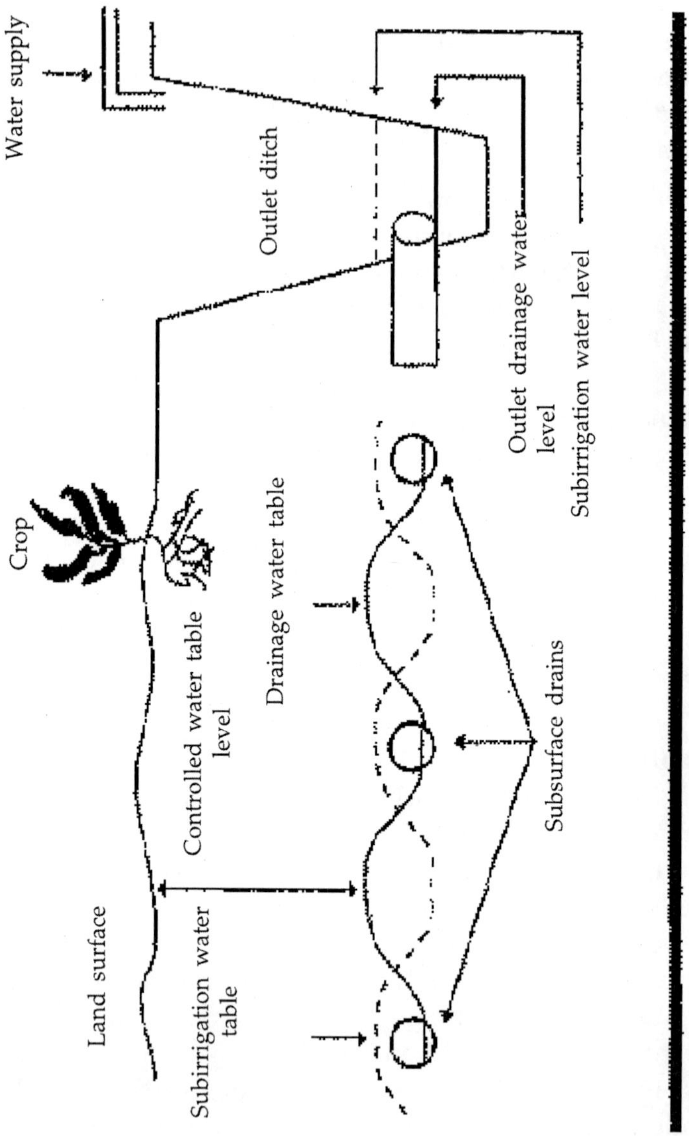

Figure 1. Water table management systems

As a result of higher water tables, water is provided for plant uptake through capillary rise or upward flux. This water helps to meet plant transpiration requirements and reduce drought stress. Corn and soybean yield increases of 10-15% have resulted from sub-irrigation in eastern Canada. An added benefit of the higher water table is that it enhances denitrification, thereby reducing nitrate leaching. Denitrification is a microbial process whereby nitrate, rather than oxygen, is used to reduce NO^{3-} to N and NO_x. Nitrates retained in the soil profile are available for plant use.

Several researchers, including Willardson et al., Bengtson et al., Gilliam and Skaggs, Evans et al. and Lalonde et al., have shown that water table management can reduce NO_3-N in drainage water by over 60%. Denitrification also occurs in water stored in ditches. Other benefits of water table management include: reduced downstream peak flows and drainage volumes, and conservation of soil water to supplement irrigation water requirements. There are, however, some instances where water table management may increase peak flows.

In areas with open canal drainage systems (with or without subsurface drainage), flashboard riser control structures may be used. These are installed in the ditches, and by either inserting or removing a flashboard, the water level in the ditch is either raised or lowered. These structures can also be used to store runoff in canals, which can later be used for sub-irrigation. Surface runoff from drained fields can also be controlled with the aid of the flashboard risers.

Water table management systems may be designed to control drainage volumes, peak flow rates or chemical losses from agricultural fields or catchments. Once the design objective has been determined, the potential hydrologic impacts and the resulting changes in water quality must be determined. Hydrologic impacts will depend on the degree and method of drainage improvement already attained in the proposed project area. High water tables will result in low infiltration capacities of the soils and high surface runoff rates, and will increase the loss of chemicals located at or near the soil surface.

On the other hand, unrestricted drain discharge (i.e., no outlet control) will produce low water tables and higher infiltration capacities. Consequently, during periods of high drain discharge, there is increased potential for the leaching of soluble chemicals existing in the soil profile. The reduction of drainage volumes may be a first step towards improved drainage water quality. Evans et al. have indicated that water table management may reduce total drainage outflow by over 30% compared to conventional drainage systems without outlet control. Outflows, however, vary widely, depending on soil type, rainfall and type of drainage system.

During very dry years, water table management may totally eliminate outflow. In wet years, it may have little or no effect on total outflow. During the growing season, water table management typically reduces outflow by less than 15 %; much of the reduction occurs during the non-growing period, i.e., winter and early spring. The decrease in outflows is not due to a significant increase in ET. A slight increase may be observed during the growing season, but this would not explain differences during the winter, when most of the reduction occurs. It is possible that most of the water is lost by lateral seepage to remote sinks, or by deep seepage to groundwater.

Depending on the topography of the watershed, this water may reappear downstream and it should be quantified. Generally, with improved drainage, there is a tendency for the planting of higher value crops, and increased chemical applications. This may lead to increased chemical loads in surface runoff and subsurface drainage discharge. Crop production methods should, therefore, focus on more efficient water, fertiliser and pesticide applications. Water quality benefits associated with water table management will only be achieved if sound agricultural practices are followed.

Nutrients

Nutrient losses in agricultural drainage water raise many environmental concerns. Eutrophication, due to high N and P concentrations, increases algal growth and reduces oxygen in surface waters. From agronomic and economic standpoints, nutrient losses represent a decrease in the efficiency of the crop production system. There are also human and animal health concerns, because nutrients, especially NO_3-N, can contaminate drinking water supplies. Concentrations exceeding 10 mg/litre can be harmful to humans.

Nitrogen, P and K are the three major elements of fertiliser and manure. Nitrogen is easily converted to NO_3-N. Table 1 shows the order of magnitude of total annual losses and peak concentrations of these three nutrients which may be expected in surface runoff and subsurface drainage. It can be seen that P and K are of less concern than NO_3-N. Nutrient concentrations in drainage water are influenced by many factors, including: rate of fertilisation; soil and crop type; and soil water regime during the growing season. These factors affect the efficiency of nutrient uptake by the crop roots, and residual nutrient levels in the soil during the fallow or dormant season.

Nitrogen

Nitrogen is transported in different forms. In most cases, water table management aims at reducing NO_3-N leaching. When developing a water table management strategy, losses of N in its different forms should be taken into account. A decrease in one form of N may result in an increase in another form. This is particularly the case if a decrease in subsurface drainage discharge results in an increase of surface runoff. In such a case, there might be an important decrease in NO_3-N concentration in drainage water, while a small increase in loss of organic N and NH_4 may occur.

Table 1. Orders of magnitude of peak concentrations and annual losses of NO_3-N, P and K in drainage water at field level

	Losses	NO_3-N	P	K
Surface runoff	Peak concentration (mg/litre)	50	10	10
	Annual (kg/ha)	1-20	1-20	1-10
Subsurface drainage	Peak concentration (mg/litre)	200	10	10
	Annual (kg/ha)	1-100	0.1-1.0	1.0-10

Knowledge of the N cycle, and the impact of agronomic practices on this cycle, will help with the implementation of a water table management strategy. Temperature and the soil water regime influence nitrification and denitrification. At the field level, the potential risks must be assessed by comparing the water table regime to:

(i) fertiliser rates and practices during the growing season;

(ii) the periods of high mineralisation rate; and

(iii) the periods of possible denitrification.

These factors are illustrated in Table 2. High water tables due to water table management may significantly decrease NO_3-N concentrations in drainage water.

Table 2. Factors influencing the nitrogen cycle in agricultural soils

Factor	Effect on crop	Effect on nitrogen cycle
Temperature	Higher ET which enhances vegetative growth and thus nitrogen uptake.	Higher temperatures (in the range of 20-30°C) increase the rate of most biochemical processes, especially mineralisation and denitrification.
Soil water regime	High water tables restrict crop growth and nitrogen uptake	Soil water contents below field capacity increase the rates of all processes, except denitrification. Saturated conditions due to higher water tables enhance denitrification.
Fertiliser and agricultural practices	Application of fertilisers increases the rate of crop growth and development.	Drainage improvements are generally linked to changes in cropping systems, which may lead to increased N fertiliser inputs. The conversion of grassland to cropland results in mineralisation of N accumulated in the soil profile.

Annual losses of total N from fields with water table management could be 40-50% lower than for conventionally drained fields. This decrease is due to both a reduction in drainage discharge and an increase in denitrification. If controlled drainage is used during the early spring or late winter, when drainage is not needed for crop production, one can, if temperatures are warm enough, dramatically increase denitrification and reduce NO_3-N losses.

The design of a water table management strategy to control N losses should take the following into account:

i. The timing and amounts of drainage discharge, before, during and after the growing season.

ii. The potential denitrification. Denitrification requires low oxygen concentration, due to shallower water tables, as well as the presence and activity of bacteria. Temperature and soluble organic carbon content in the soil influence denitrification. Soil temperatures above 10°C are necessary to induce significant denitrification.

iii. N_2O is a by-product of denitrification. This is a greenhouse gas, which is of environmental concern, and may only be a problem on very large watersheds.

iv. The scale at which the water table management is implemented. At the watershed scale, there is the possibility of less vegetation in and along channels. This may yield higher NO_3-N levels at the field edge.

v. Continually high water levels in drainage canals and/or ditches can cause problems of ditch bank stability and also promote problems with beavers and other rodents.

Phosphorus

Phosphorus losses are closely tied to sediment loss. Surface runoff is, therefore, the primary factor influencing both sediment and P transport. Shallow water tables during high rainfalls could increase surface runoff, and hence P movement. However, at the watershed scale, controlled drainage has been shown to reduce P losses.

Pesticides

Pesticides are generally transported with soil particles in surface runoff. According to Munster et al., pesticide concentrations are generally higher in surface runoff than in subsurface drainage water. However, Kalita et al. stated that high pesticide concentrations in subsurface drainage water could be expected if preferential flow occurs. The amount of pesticide in subsurface drainage water is typically less than 0.1% of the amount applied. The pesticide most studied is atrazine, or compounds of the same family. This is a herbicide for maize crops.

Results for aldicarb, alachlor and metolachlor are also available. The concentration of these chemicals is generally very low, in the range of 1-3 m g/litre. Although further investigation is needed, a water table management strategy that aims to reduce pesticide losses should preferably reduce the amount of surface runoff. The water table level should therefore be lowered if a rainfall occurs immediately after pesticide application. However, Michaelsen showed preferential transport of several herbicides being enhanced in drained plots in northern Germany. Most pesticides have a relatively short half-life (measured in days), and therefore do not persist in the soil or water for long periods.

Operation of Water Table Management Systems

Water table management systems can be operated by farmers, water boards or water users associations. However, there is also a need for an institution to monitor water quality. Sometimes a compromise may be required between agricultural and water quality benefits.

Depending on the type and severity of water quality constraints, water table management will seek an optimum for either agricultural production or environmental protection. System constraints may be either physical, technical or human. The most critical physical aspects include topography, the existing drainage network, water availability (for sub-irrigation) and soil conditions.

Water table management requires greater farmer involvement than does conventional drainage. The operation of subsurface drainage systems requires few management decisions. However, with water table management, operational indicators are difficult to see, as the response to outlet adjustments is not immediately evident in the field. Furthermore, the operational incentive may not be solely increased crop yield, but also improved water quality and increased water quantities for downstream water users. These factors must be clearly defined for system operators at the start of the project.

There are two approaches to the implementation of water table management: the catchment level (or large tracts of land) and the field scale. These two approaches are complementary. The scale of implementation depends on: water quantity and quality objectives; farming systems; regional hydrology; institutional arrangements; and topography. Water quality is linked to farming practices and is therefore to be controlled primarily at the field level. However, field systems may necessitate control structures at the catchment level where water quality control depends on a reduction of drainage volumes and sediment transport.

Location of the control structures within the catchment will depend on the discharge and drainage area above the project site. The upstream drainage area may be several thousand hectares. In this case, it may be more economical to control the water level in field ditches with smaller structures. This may cost less than larger structures on main channels. The existing drainage network is also an important consideration. The depth and width of the channels govern water storage capacity. If a dual controlled drainage/sub-irrigation system is to be implemented, channel storage capacity has to be sufficiently large to provide irrigation water.

Water table management is best suited to relatively flat lands. Sloping lands would require many control structures, to provide uniform water tables. Land slopes of less than 1% are recommended. However, as pointed out by Evans and Skaggs, while there is no absolute limit, slopes of less than 0.1% are the most practical. At the watershed scale, a slope of 0.1% would require a control structure every 300-1 000 m of channel length. Soil permeability is a key factor governing the rate of water movement from ditches to adjacent fields, and the upward movement of water from the water table to the plant roots.

Good lateral movement of water will occur in moderate to highly permeable soils. While Evans and Skaggs recommended a permeability of about 0.45 m/day, experience in eastern Canada has shown that water table management is also feasible on soils with permeabilities of 0.3 m/day. A restrictive soil layer, not far from the bottom of the subsurface drain pipes, will reduce deep seepage losses.

Generally, a monitoring system needs to be implemented, both for system management and to measure water quality changes. Monitoring involves several factors and has to be performed on a regular basis, both at the field level by the farmer and at the catchment level by the water authorities.

Actual and potential detrimental impacts of drainage water quality both within the watershed and downstream should be identified. These investigations will help determine the principal water quality constraints and the most sensitive periods for control. They are best conducted by water authorities. The data may be useful in guiding fertiliser application.

Water table management requires periodic monitoring of the water table at the midpoint between drain laterals or ditches. It is best done by the farmer. One observation well per field is the minimum recommended. For a given soil and drain spacing, the frequency of observation and adjustment of the control structure depend mainly on the weather and crop development stage. For example, when crops are in their early stages of development, a shallow water table may impede proper root development, and this could make the crop more susceptible to drought later in the year.

Table 3. Monitoring requirements

Impacts	Parameters to monitor	Monitoring interval
Peak flows	Surface runoff, subsurface drain discharge, water levels in streams	1-hour intervals with automatic devices
Soil erosion	Sediment in surface runoff	Accumulation after a rainstorm, using sediment samplers
Nutrient losses	Nitrate, ammonium, phosphorus	Monthly, during the growing season
Pesticide losses	Pesticide concentrations	Monthly, during the growing season

The response of the water table to rainfall and control structure adjustments may be slow, depending on the soil properties and drain spacing. It may take several years to fully understand the response patterns and successfully operate the system. Accurate weather forecasts and the ability of the farmer to make use of climatic data could improve system operation.

Water agencies should organise a medium or long-term monitoring programme of the impacts of water table management on the hydrology, water quality and crop performance in the project area. Some of the main parameters to be monitored are shown in Table 3.

5

Drainage Water Quality

Characteristics of Water Quality

Water quality is relative and is defined as the characteristic of a water that influences its suitability for a specific use. Quality is defined in terms of physical, chemical and biological characteristics. Drainage water is no different from any other water supply and is always usable for some purpose within certain quality ranges. Beyond these limits, drainage water must be disposed of in a manner that safeguards the usability or quality of the receiving water for present established and potential uses. Surface and subsurface drainage water from irrigated agriculture is normally degraded compared with the quality of the original water supply.

Drainage water that flows over or through the soil will pick up a variety of dissolved and suspended substances including salts, organic compounds and soil particles. Management for safe re-use and disposal requires an understanding of the characteristics of the drainage water, and a matching of those characteristics to the environmental protection needs of the re-use or disposal area. Both surface and subsurface drainage effluent contains substances that are potential pollutants. These pollutants may be:

— purposely introduced into irrigation water;
— mobilised by the practice of irrigation and/or drainage; or
— concentrated as a result of ET.

Each of these three processes requires a different management approach to mitigate its impact. For example, in the first instance, the pollutant source could be removed or limited. By contrast, the concentration of dissolved solids by ET is a natural process that results from consumptive water use and is unavoidable. One or more of these processes may be occurring, thus increasing the complexity of assessing the problem and developing solutions to a given water quality problem.

Pesticides

Numerous types of pesticides may occur in drainage water. This makes it difficult to assess their potential impacts on water quality. Most pesticides are synthetic organic compounds and there are documented instances where organic pesticides in irrigation runoff have caused downstream water quality problems. Recent studies in both the San Joaquin and Imperial valleys of California have shown that surface runoff carries pesticides that cause toxicity problems in aquatic organisms in surface streams. These pesticide problems are the result of farming practices, not the design or functioning of the drainage system.

Drainage discharges from irrigation may present elevated concentrations of pesticides, but only for a relatively short period of time. The best solution is improved irrigation water and pesticide management. The drainage system should not be considered a mechanism for controlling pesticide runoff. Those designing and building a surface water drainage re-use or disposal system need to consider whether a pesticide runoff problem is likely to occur, the type of pesticides expected, and the mitigation steps needed to avoid downstream pollution. High pesticide concentrations in subsurface drainage water are less probable because of the filtering action of the soil.

Few recent surveys of subsurface drainage water have been conducted, but groundwater surveys in California show few pesticide detections. Because of infrequent detections, pesticides in subsurface drainage water should not be considered a priority for intensive evaluation unless general water quality surveys show detected levels of concern. However, data summarised by MacKenzie and Viets show that pesticides have been found in subsurface drainage systems. In the event that pesticides are found, the source is probably a farming practice which can be altered to prevent the continued deep leaching of the pesticide. The presence of a pesticide in subsurface drainage water indicates that the pesticide is in the shallow groundwater and will continue to be discharged with the subsurface drainage for a long period.

Trace Elements

Inorganic trace elements are different from synthetic organic compounds (pesticides) in that they are commonly present at low levels in nature and there is already a natural level of tolerance. There is, however, a fine division between natural tolerance and toxicity. It is therefore essential to have good information on the concentration of trace elements in the drainage water in order to develop safe re-use and disposal methods. High concentrations of inorganic trace elements in irrigated soils and groundwater pose a threat to the environment if they are mobilised by irrigation and drainage practices. They can be collected in the drainage water and then discharged at relatively high concentrations into the environment or at low concentrations and bioaccumulated in the food chain. These trace elements pose a threat to agriculture, wildlife, drinking water and human health.

The trace elements of most importance are those designated as priority pollutants by the USEPA and documented as pollutants associated with irrigated agriculture. Deverel and Fujii divided these priority pollutants into four categories (Table 1) and reviewed the factors influencing their occurrence and mobility in irrigated soils and groundwater. Recent surveys of subsurface drainage water in the San Joaquin, Sacramento and Imperial valleys of California, have shown that trace element concentrations can be elevated and appear to be strongly associated with the geologic setting of the irrigated area.

Table 1. Geochemical behaviour of trace elements

Alkali and alkali earth metals	Transition metals	Non-metals	Heavy metals
Barium	Chromium	Arsenic	Cadmium
Lithium	Molybdenum	Boron	Copper
	Vanadium	Selenium	Lead
			Mercury
			Nickel
			Zinc

Deverel and Fujii also concluded that high concentrations of trace elements in soils and groundwater in arid irrigated areas can occur concomitantly with high soil and groundwater salinity and be affected by the same processes that affect the soil and groundwater salinity. They concluded that high trace element concentrations may also occur independently of salinity such as in humid area soils. Therefore, trace elements should be a priority for evaluation in all drainage projects.

There are no data to show that trace element levels are higher in the surface drainage flow than in the applied water before it moves across a field. Increases in trace element concentrations in surface runoff are generally not expected. Therefore, they are not considered a priority for evaluation. It is important to be able to predict the potential occurrence of high trace element concentrations in subsurface drainage water. Bowen and Shacklette and Boerngen developed range and geometric mean values in soils for the priority pollutant trace elements.

Comparison of actual soil data with these background levels would give an initial screening as to the potential for trace element leaching. This evaluation should be conducted in conjunction with shallow groundwater sampling for the same potential elements of concern.

Nutrients

The two major nutrients in drainage water are N and P. Both contribute to eutrophication of surface waters. Nitrogen can be in either the organic form (ammonium) or the inorganic form (nitrate). The predominant form in surface drainage is organic N. This comes from the organic matter being washed off the field and should be the focus of any water quality evaluation for surface drainage. This is a persistent problem in higher rainfall areas but is not usually an issue in arid areas. Ammonia is adsorbed on clay particles due to their positive charge. It can also volatilise.

Nitrate is the dominant form of N in subsurface drainage water. Nitrate N should be the focus of any evaluation of water quality from a subsurface drainage system. High nitrate concentrations in subsurface drainage can originate from a number of sources: geologic deposits, natural organic matter decomposition and deep percolation of nitrate resulting from fertiliser applications. Nitrate contamination of subsurface drainage water has been documented by Madramootoo et al.. Mineral N, in the form of nitrate or nitrite, is transported in a dissolved form. The proportion of these various forms in drainage water depends on the predominant form of drainage.

Nitrite is the most dangerous form of N. It is in general a transient form of N, present in small quantities. Agricultural drainage water also contains phosphate in both organic and inorganic forms. Most of the phosphate in surface drainage is in the organic form. Little phosphate has been found in subsurface drainage water because of its strong adsorption in arid zone soils and in humid regions.

Sediment

While rainfall induced erosion is a significant source of sediment, irrigated agriculture may also cause erosion directly through application of irrigation water, or indirectly through sub-optimal land management. Sediment contained in surface runoff from agricultural lands may carry P and certain pesticides to surface waters where they may contaminate the food chain or affect other beneficial uses of water. Excess sedimentation can also degrade the stream environment, diminish the health and diversity of fish and wildlife habitats, limit recreational uses and add to the costs of flood management and drinking water treatment.

Because subsurface drainage water is primarily groundwater, it is not expected to carry significant amounts of sediment. No data are readily available on sediment loads in subsurface drainage water, but sediment from subsurface drains is not normally expected to be a significant issue. However, from time to time a subsurface drain line will be poorly constructed, or a drain line will fail after construction and a significant amount of sediment will collect in the drains. This will then either plug the drains so they will not function or be discharged at the drain outlet.

Moreover, in certain soils, drains with inadequate or improperly installed envelopes can accumulate fine sands and silts and these sediments will eventually either plug the drains or be discharged with the drainage water. In any event, the normally relatively sediment free water from subsurface drains might erode the banks of unlined surface drains and increase the sediment load of the drainage water. Erosion is a natural geologic process. However, although sediment cannot be eliminated, its movement can be controlled to reasonable levels.

Bacteria

Bacteria are a potential pollutant where surface return flows come from land that has received applications of human or animal waste. Bacterial pollution may also originate from wetland discharges. The focus of a bacteriological contamination assessment is on the measurement of coliform and faecal coliform levels in the water. A normal irrigated farming operation would not be expected to produce adverse bacteriological levels in surface drainage water.

The presence of coliform or faecal coliform would be an indication that other types of wastewater, such as municipal, industrial or animal waste, may have entered the surface drainage system. Soil is a biological filter. Therefore, it is not expected that micro-organisms will move through the soil from surface water to a subsurface drainage system. There is no record of coliform or faecal coliform levels being measured in a subsurface drainage system, nor are they expected to be a problem.

Temperature

Elevated temperatures occur where irrigated fields or wetlands are warmed by the sun, and tailwater from these areas is then discharged into a stream, causing a rise in the stream temperature. This problem is often aggravated where diversions for irrigation and wetland management also reduce the total stream flow. Elevated temperatures have a direct impact on stream aquatic life especially in certain cold water streams or those with anadromous fisheries.

Temperature surveys should be an essential component of any surface water monitoring if elevated water temperatures are expected. Power plants sometimes affect the temperature of downstream waters. In contrast to surface drainage, subsurface drainage is mainly groundwater and has a relatively constant water temperature. The temperature of subsurface drainage waters can be easily predicted as they approach average seasonal or annual temperatures. Measurements in the San Joaquin Valley of California, show that subsurface drainage discharges average 17°C, with a variability of less than 2°C. These cool discharge temperatures would not be expected to have an impact on warm water fisheries as they quickly come into equilibrium with the ambient air temperature.

Salinity

The practice of irrigation results in consumptive uses of water, leaving behind salts concentrated in a smaller volume of water. During irrigation, two types of drainage may occur: surface drainage and subsurface drainage. Reeve et al. showed that the salt content of water changes little in flowing over soil even where there is a visible salt crust. Salts are a major water quality factor in choosing disposal options for subsurface drainage in arid irrigated areas. Salinity can restrict the urban or agricultural re-use of drainage water, as it is the most significant long-term water quality concern for managing irrigated agriculture in arid zones.

Salinity has not been noted as a serious concern with subsurface drainage waters from humid areas. This is generally due to the higher rainfall, higher dilution capacity in surface waters and lower initial salt content in the soil. In arid irrigated areas, irrigation practices mobilise naturally occurring salt in the soil and concentrate those salts already present in the supply water. The salts captured by a subsurface drainage system are often highly concentrated with the major cations being Na^+, Ca^{2+}, Mg^{2+} and, to a lesser extent, K^+. The major anions are Cl^-, SO_4^{2-}, HCO^{3-}, NO^{3-} and CO_3^{2-}.

Most subsurface drainage waters are $NaSO_4$ and NaCl dominated. These are common non-toxic elements that only become problematic when concentrated in the soil. However, some $MgSO_4$ waters have been observed. The predominant cations and anions influence the re-use and disposal methods chosen. Subsurface drainage water from arid areas always has a higher salinity than the supply water, a higher proportion of Na and Cl, an increased hardness and a higher sodium adsorption ratio (SAR).

The higher salinity and higher levels of specific ions often reflect the characteristics of the soil through which the irrigation water has percolated. This in turn is influenced by the shallow groundwater quality, by the ionic composition of the irrigation water, and by the irrigation efficiency. Salt species are also influenced by a number of interdependent, multi-phase chemical interactions. There are a number of models available to predict solute migration in soils. A full salinity appraisal is an essential component of any subsurface drainage water re-use or disposal scheme.

Sulphurous Compounds

Water quality problems are often encountered when sulphaquepts or acid sulphate soils are drained. Such soils are mostly found in tropical river deltas. Improved drainage of sulphaquepts results in an increased discharge of acid water. High quantities of sulphuric acid may be released into receiving streams. There is also a reduced pH which can affect aquatic life. Furthermore, iron and aluminium in these soils can be mobilised and may cause human health problems if the receiving water bodies are used for domestic drinking water.

Protection of Water Resources

The re-use and disposal of agricultural drainage water must be environmentally sustainable. This means allowing other uses to be made of the water resources that will not be adversely affected by the drainage project. Understanding the water quality needs of the various downstream uses can assist in developing mitigation methods. The primary needs to be considered are drinking water, industrial supply water, agricultural supply water, recreational uses and aquatic life.

Drinking Water

Surface and subsurface drainage water discharges can degrade the potability of drinking water and increase the cost of testing, treatment and delivery. An additional concern is human contact with the water during bathing, washing clothes and recreational use. Domestic use and drinking water safety is a primary consideration for most nations and for the general public. Drinking water supplies can be affected by sediment, nutrients, micro-organisms, total salts, sulphate, chloride and pesticide levels. Sediment increases water treatment costs. High nutrient levels, especially nitrate above 45 mg/litre as nitrate, are a potential public health problem.

High nutrient levels may also cause algae problems which can lead to a disagreeable taste and the formation of THM precursors. Bacteria levels indicate a high potential for disease problems and may increase water treatment costs. High total dissolved solids (TDS) can cause taste problems. Recommended levels of salinity are less than 0.9 dS/m with a maximum suggested at 2.0 dS/m. Sulphate and chloride levels are recommended between 250 and 500 mg/litre.

The higher salinity of subsurface drainage water may also increase treatment costs due to the increased water hardness. Pesticide levels in drainage water should be checked if a discharge is planned into a known drinking water supply. Trace elements may also cause toxicity.

Water Supply for Industrial Use

Water supplies for industrial processing are affected primarily by high salinity levels. The major concern is increased hardness as this could cause scaling or precipitation and increase costs for treatment prior to use. The discharge of drainage water into a known industrial supply should focus on the salinity and hardness limits of the water. Secondary effects of drainage waters on industrial usage might be related to nutrients, sediment, micro-organisms and pesticides. These may restrict the type of use the water is put to, or may increase treatment costs prior to use.

Agricultural Supply

Agricultural water users are primarily affected by salinity or the individual salts in the water supply. Total salinity can cause crop yield reductions and increased costs for salt management. Specific salts such as boron, chloride and sodium also cause plant toxicity problems. In addition, sodium can cause soil permeability problems. As subsurface drainage water itself is commonly higher in total salts, sodium, chloride and boron, agricultural water users can be affected by its discharge into usable water supplies. There are several published guidelines on agricultural use of higher salinity water.

Trace elements are of concern for irrigation or animal drinking water use. Animal drinking water guidelines are given in Ayers and Westcot along with a discussion of the guideline values for irrigation. In most instances, trace element levels acceptable to irrigated agriculture may be lower than those required for other uses, such as aquatic life, so agricultural use may not be the most limiting. The exceptions are molybdenum and selenium which can accumulate in crops and cause toxicity to the consumer of the crop.

Protection of Aquatic and Wetland Wildlife

Public values for the protection of aquatic and wetland wildlife are concerned with the survival, health and diversity of fish, wildlife and aquatic plants. If discharged incorrectly, pollutants in subsurface drainage may be acutely toxic or bioaccumulate. They may also damage plant and animal health in the long term, or impair the reproductive ability of aquatic species. The pollutants of primary concern in surface drainage water are pesticides, sediment and temperature.

Elevated temperatures can cause damage to cold water fisheries. Sediment levels can cause instream damage to habitat or restrict light penetration and aquatic food development. With surface drainage, the pollutant of greatest concern is pesticides. However, there are few criteria available on allowable pesticide levels in water. It is therefore suggested that pesticide surveys and analysis concentrate on aquatic toxicity.

The pollutants of primary concern in subsurface drainage are trace elements. In the drainage water discharged areas, there are six potentially toxic trace elements known to exist in concentrations sufficient to cause impacts on aquatic life: As, B, Hg, Mo, Se and U. Mercury and Se are of particular concern for aquatic life because of their bioaccumulative nature, even at very low concentrations in the water. Boron is known to be harmful to agricultural crops and similar impacts on aquatic plants would be expected. Arsenic is a known human carcinogen at very low levels and could have a similar impact on aquatic life and wildlife, though no data are available.

Plants tolerate high levels of Se and Mo, but both elements are known to concentrate in forage crops and can have an impact on the livestock or wildlife that consume the crop. Little or no data are available for U. Trace elements such as Cu, Fe, Mn and Zn have little solubility in saline and alkaline soils and would not be expected to be present under such conditions.

Recreation

Recreation in water is becoming an important part of many economies. The discharge of subsurface drainage water could affect the aesthetic or recreational value of a water body and impair that component of the economy. The impairment would be to restrict human contact with the drainage water due to high pesticide, sediment or bacteria levels. These impacts would generally come from surface drainage systems. A secondary concern with recreation and associated with subsurface drainage water is the discharge of nutrients that may cause eutrophication and stimulate aquatic plant growth. This leads to a loss of aesthetic values in recreational waters.

6

Drainage Water Treatment

Selection of Treatment Process

The first steps in the selection of any treatment process for improving drainage water quality are to thoroughly define the problem and to determine what the treatment process is to achieve. In most cases, either regulatory requirements or the desire to re-use the water will be the driving force in defining the treatment issue(s) to be selected for a particular drainage water. A thorough knowledge and understanding of these water quality criteria is required prior to selecting any particular treatment process. Most of the treatments discussed will not reduce the salt concentration in the water and some may result in increased salt concentration.

Historical data of the chemical constituents in the drainage water are used for decision making. Once the framework that governs the selection of the treatment process has been identified (e.g., regulatory requirements, re-use, or both), then the available data must be evaluated as to its adequacy for the preliminary selection of possible treatment options. In general, a more complete data record, in terms of both the length of record and the constituents monitored, will make the task of treatment selection easier. As a minimum, one would need the ranges of expected flow volume and concentrations of the constituents of concern. In addition, knowledge of any seasonal variations of flow or constituents would be highly desirable.

If data are not available, a portion of the project budget should be allotted to obtain this information in order to avoid costly mistakes at a later stage in the process. It is also necessary, when using historical data, to establish the reliability of the data. This can be done by reviewing the analytical procedures used, by comparing data with an independent source, or by comparing historical values with more recent data. The next step in developing a successful treatment programme is to identify the selection criteria. This will require input from all parties involved in the project including the funding source(s), users (farmers, industries or municipalities), water districts, regulatory agencies and the public.

Criteria that will affect process selection include: capital cost, O&M costs, land requirements, level of treatment required and regulatory constraints. These criteria need to be defined as closely as possible in the earliest stages of process selection. However, it is important to note that there will undoubtedly be modifications to the criteria as the project progresses. Once the selection criteria have been defined, possible treatment options can be identified.

A good way to begin is with one or more 'brainstorming' sessions where any possible options are first identified without any suggestion being rejected. It is important to have input from individuals with experience and expertise in a variety of areas, so as not to 'zero-in' on a particular type of technology initially. This list of all possible options should then be evaluated against the identified selection criteria and ranked according to how each option meets the criteria. It is also appropriate to evaluate the known advantages and disadvantages of each option. From these evaluations, there may be one clear option to pursue or there may be several options which meet the selection criteria. If there are several options, then additional, more stringent criteria should be used to narrow the field down to two or three preliminary selections.

Unless the preliminary option or options are standard processes in use with drainage water or have been successfully used in applications with water very similar to the water in question, it is highly advisable to conduct pilot-scale tests. In general, the larger the pilot system, the better it will indicate the efficiency of the process when built at full scale. The single most important issue in pilot-scale operation is to use the water which will actually be treated by the full-scale process. Drainage water is such a complex solution that synthetic mixes, while useful for initial and fundamental laboratory studies, are poor substitutes for evaluating actual treatment processes.

Conditions in the pilot-scale system should approximate as closely as possible the operating conditions of a full-scale system. Hydraulic retention time (residence time), flow velocities, applicable loading rates and removal efficiencies of target constituents are typical parameters of concern to be measured and evaluated in the pilot testing. If the pilot tests do not show that the treatment process can achieve the desired results, then the process must be modified and retested or other options must be investigated. Once pilot testing is complete and there is at least one process that has been shown to be effective, then the best option can be selected for full-scale implementation.

The options should be judged against the selection criteria developed early in the process. At this point, other factors concerning the processes should also be evaluated such as: complexity; number and expertise of personnel required to operate and maintain the system; waste products produced and their disposal; expandability of the system; ability of the system to remove the target constituents to below the currently desired levels (e.g., in case regulations become more stringent); and ability of the system to remove constituents which are not currently of concern (e.g., in case new constituents become regulated). The consensus of as many of the involved parties as possible is desirable in the final selection of the best treatment option. Achieving consensus at this point will avoid conflict in the future.

Treatment Methods

Treatment approaches can be divided into three general types: physical, chemical and biological. Many processes exhibit both physical and chemical aspects and so are sometimes called physical/chemical or physicochemical treatment.

Physical Treatment

Particle removal

Several physical processes aim to remove suspended particulate matter. While subsurface drainage water itself is usually low in suspended particles, these processes might be used in an overall treatment process for the removal of particulates formed in other stages of the treatment, such as removal of bacteria from a biological system or removal of precipitates formed in a chemical treatment process. Particle removal unit processes include sedimentation, flotation, centrifugation and filtration. Filtration further includes granular media beds, vacuum filters, belt presses and filter presses.

Adsorption

Adsorption is the process of removing soluble contaminants by attachment to a solid. A common example is the removal of soluble organic compounds via adsorption onto granular activated carbon (GAC). GAC is useful for its ability to remove a wide range of contaminants. Certainly, if pesticides were a concern for the drainage water being examined, the use of GAC adsorption would be a leading candidate for treatment.

Air stripping

Another possible treatment for removing volatile compounds from water is air stripping. In a conventional countercurrent air stripping operation, the contaminated water is distributed at the top of a tall reactor vessel that is packed with materials or structures with a high surface area and void ratio. As the water advances downward, clean air is introduced at the bottom of the reactor and moves upward. As the water and air make contact, volatile compounds are transferred from the liquid phase to the gas phase according to gas transfer theory.

Membrane processes

If removal of salts and production of a high quality water is the treatment objective, membrane processes or distillation will be leading process candidates. The separation of salts and organic compounds can be accomplished by using a selectively permeable membrane. Membrane processes are also finding use in the water treatment industry for removing particulates and microbial contaminants such as Giardia and Cryptosporidium.

Membrane processes can be divided into three main categories: dialysis, electrodialysis and reverse osmosis. Each of these processes requires some type of driving force energy to separate the contaminants from the clean water. For dialysis, the driving force is the difference in concentrations of the contaminant across the membrane (a concentration gradient). For electrodialysis, the driving force is an electrical potential. For reverse osmosis, the driving force is applied pressure.

The use of membrane processes for salt removal has generally been considered too expensive for drainage water, but new developments in membrane technology may make this an attractive option for treating at least a portion of the flow to reduce the total salt concentration. The use of a membrane process such as microfiltration for particle removal may also be a cost-effective alternative as part of an overall treatment system.

Distillation

Distillation is a thermal process used for salt removal. Heat is used to vaporize the water, leaving the salts behind. The water vapour is condensed to a high quality water. Distillation is energy intensive and has largely been replaced by reverse osmosis for desalination applications.

Coagulation and flocculation

Coagulation and flocculation are used to remove particles of all types from water. The particles might be colloids present in the drainage water that are too small to remove by gravity settling or filtration, or they might be colloidal precipitates formed during a treatment process. Coagulation is the process by which these small particles are destabilised and the initial aggregation of the destabilised particles into larger particles called floes. Coagulation is accomplished by the addition of a coagulant, which can be either an inorganic metallic salt such as alum (aluminium sulphate) or ferric chloride, or a high molecular weight organic polymer.

The coagulant serves to neutralise interparticle charge repulsion and to enmesh the particles into an aggregated floe. Flocculation is a slow mixing of the particles to bring them into contact with one another to form even larger particles. The objective is to produce a large fast-settling floe.

Chemical Precipitation

Certain compounds, especially metals, can be removed by changing their solubility to cause their precipitation. Many metals can be precipitated as a metal hydroxide by increasing pH with lime or caustic soda to achieve the pH of minimum solubility. These would include Cr, Ni, Cu, Fe, Pb and Hg, and other elements such as As. The pH of minimum solubility varies according to the metal in question. Precipitation can also occur by the formation of insoluble compounds through adding certain chemicals, or from chemical species formed during another treatment step. Such is the case with sulphides and carbonates, which can be formed during biological treatment and combine with metals and some cations to form precipitates. Precipitation can lead to a slight net decrease in TDS.

Exchange of ion

Ion exchange involves the chemical exchange of ions

(charged dissolved molecules or atoms) in solution with ions on a solid phase. The solid phase, usually a synthetic organic resin, is chosen to specifically adsorb the constituent(s) of interest. To maintain the electrical charge balance, the resin must release an equal amount of charge into solution. For example, in water softening, a cation exchange resin initially holds Na^+, but releases them as it adsorbs Ca^{2+} and Mg^{2+} from the water stream moving through the resin.

Anion exchange resins are also available. For either type of resin, there is a preferential adsorption of either the cations or the anions if more than one type is present. For example, calcium is more strongly adsorbed than magnesium, and sulphate is more strongly adsorbed than nitrate. Certain resins have been developed to maximise their affinity for a specific ion (e.g., borate).

Advanced oxidation processes (AOPs)

One major disadvantage of both GAC adsorption and air stripping for organic compounds is that they only transfer the contaminant from one phase (water) to another phase (carbon or air, respectively). Advanced oxidation processes (AOPs) are capable of compound destruction, or more accurately, the mineralisation of chlorinated organic compounds to non-toxic constituents such as carbon dioxide, water and chlorides.

AOPs rely on the production of highly reactive radicals to break down the organic compounds. AOPs are commonly based on the use of hydrogen peroxide (H_2O_2) or ozone (O_3) in combination with ultraviolet (UV) light to cause radical formation. Another type of AOP uses photoactive metal catalysts and UV light to generate the radicals. Their high cost is the primary disadvantage of most AOPs.

Biological Treatment

Biological treatment can be a useful tool in drainage water treatment for the removal of both organic and inorganic contaminants. Biological treatment usually refers to the use of bacteria in engineered reactor systems for effecting the removal or change of certain constituents, such as organic compounds, trace elements and nutrients. Algae have also been used and natural wetlands systems can be used in some cases to replace conventional reactors. The bacterial reactions involved can be divided into two major categories according to the use of oxygen by the bacteria.

In aerobic systems, O_2 is provided and used by the bacteria to biochemically oxidise organic compounds to carbon dioxide and water, and possibly to oxidise reduced compounds before their release to the environment. Aerobic systems are usually odour free. In an aerobic system, oxygen is the electron acceptor and organic carbon sources are usually the electron donors in the biochemical reactions that take place. In an anaerobic system, oxygen is excluded and the bacteria utilise compounds other than molecular oxygen for the completion of metabolic processes.

Biological reactor types can be broadly divided into two types: suspended growth and attached growth. In suspended growth systems, the bacteria are grown and maintained in suspension by mixing of the liquid contents. In attached growth systems, the bacteria grow in a thin layer (a biofilm) on an inert support, such as plastic media or fluidised sand. Both aerobic and anaerobic systems can have either suspended or attached growth depending upon the reactor configuration.

Treatment in Constructed Wetlands

The protection of wetlands is one area of public concern

where agricultural and environmental values need to be further harmonised, shared and promoted. Ecologically sensitive wetlands need to be distinguished from poorly drained, wet, depressional cropland. The latter is farmland whose productivity can only be improved through better drainage. Wetlands which are transitional zones between terrestrial and aquatic systems serve valuable ecological functions. The water table is usually near the surface, and such wetlands support fish, wildlife, waterfowl and aquatic plants. They also trap sediment and pollutants, and cycle nutrients.

Wetlands are of hydrologic value in that they serve as flood control systems, recharge groundwater, and maintain instream flows. They also have economic value if they support recreation and other economic functions. A challenge is to ensure that agricultural and drainage activities do not negate the environmental value of wetlands. Cropland and wetlands must be viewed as part of a viable and sustainable ecosystem. Constructed wetlands for domestic and industrial wastewater treatment are gaining acceptance in many countries. They can also be beneficial in providing treatment of agricultural drainage wastewater under many circumstances.

Flow-through wetlands are one option for the management of agricultural drainage water. They can be described as constructed wetlands or natural depressional areas prepared to facilitate the movement of surface waters through specially selected vegetation. Flow-through wetlands are an alternative to evaporation ponds, and should be considered. Where the water quality is appropriate, they are preferable to evaporation ponds. Evaporation tends to concentrate the ions contained in the drainage water, and some of these substances can become toxic in the ecosystem.

Concentration of the ions reduces the opportunity for beneficial use of the drainage water in downstream areas by degrading receiving waterways or water bodies. Wetlands can reduce some pollutants such as N, P and sediment, and improve water quality. Wetlands consume water and will reduce the total available water for downstream users even though the general water quality may be improved. Water use by wetlands is approximately equal to potential ET rates. The primary values of flow-through wetlands for drainage water quality improvement from irrigated areas are to:

- — reduce the evaporation opportunity time which will minimise the direct concentration of salts and minor constituents which may become toxic when concentrated;
- — provide physical filtration and sedimentation of soil particles and attached contaminants; and
- — provide vegetation to remove excess N, P, K and organic wastes.

Secondary values in the use of flow-through wetlands are for wildlife habitat, recharge of groundwater aquifers, wind erosion control, and fish or shellfish production by:

- — providing breeding, nesting, feeding and cover habitats for invertebrates, insects amphibians, reptiles, birds and mammals;
- — providing water detention that will facilitate infiltration of surface waters through the soil profile with opportunity for degradation of chemicals and vegetative uptake of pollutants;
- — reducing wind erosion in the vicinity of wetlands, thus preventing wind blown contaminants in the air; and

— providing potential for fish or shellfish production if the pollutants present will not cause health problems to consumers.

Site selection, data collection and the operational needs are of critical importance when planning and designing flow-through wetlands for drainage water management. Wetlands are often created to provide waterfowl habitat. However, consideration must also be given to the physical benefits achievable if the wetland or series of wetlands are developed to improve the quality of drainage waters. The topographic, soil and geologic characteristics will have a great impact on the effectiveness of pollution control benefits from the completed system.

The vegetative materials selected for growth within the wetland will determine the success of the effort. The vegetation must be tolerant to the extremes in salinity and pollution anticipated and must be selected to maximise the uptake of nutrients and pollutants. Normally, a series of wetlands with different vegetative and soil materials will facilitate operational management and improve system effectiveness. Considerations for harvesting the vegetation and management of the vegetation may require interior dikes and gate structures.

To provide practical removal of some pollutants, harvesting of the vegetation will be required and economic considerations for disposal or marketing of the vegetation should be evaluated. Plants are being developed that have a superior capability to remove pollutants from soil, such as penny cress and other metal 'scavengers'. These plants can be grown periodically to rejuvenate the wetland and enhance the sustainability of treatment benefits. At the end of the growing season, the plants can be cut, dried and burned. Metals can be extracted from the ashes for recycling.

Site selection involves consideration of hydraulic and geohydraulic characteristics of the soil and subsoil to ensure the success of the system. Flow-through wetlands must be located where the drainage waters to be managed can physically be delivered. The topography must be suitable to allow the construction of dikes and outlet facilities. Proper soil must also be available nearby for embankments and lining the wetted perimeter of embankment areas. The site must have depressional characteristics that are large enough to allow the creation of water bodies with a depth range that will facilitate a vigorous growth of the selected vegetative materials. Water depths in the wetlands must be controlled to provide the appropriate environment for growing the specific vegetation desired.

The water depth and flow regime must also consider vector control and health considerations related to water borne diseases. The sizing of each wetland should be related to the opportunity time required for the uptake of the pollutant or pollutants that each vegetative wetland area is designed to reduce. The sizing of the constructed wetlands should also consider the volume of drainage water that will be transported to the wetland system for improving its quality. The water discharged from the wetlands must have a suitable outlet.

Effective constructed wetlands should permit vegetative materials to proliferate. Soils of the wetland should have a permeability which is sufficiently low to minimise percolation. Wetlands should be located in areas that are not porous, as retention of water is the primary characteristic required. If the site has sections of porous soil, then borrow pits with clays or other less porous material may be required to facilitate the lining of these highly permeable areas. As pollutants are removed in the initial wetlands, drainage water is discharged through lower wetlands in the system and it may recharge the groundwater system in the lower areas.

Barriers to infiltration thus become an important feature in the design of systems in the upper wetlands and are not necessary in the lower wetlands. The accumulation of dead plant material, suspended solids, algae and other clogging material on the bottom can significantly reduce seepage rates. Cavernous geologic areas should be avoided unless special precautions for seepage control and dike stability are taken. The expected quality of infiltrated waters from a constructed wetland system and the groundwater must be analysed to ensure that groundwater degradation does not occur.

As far as is both possible and practical, the vegetative types must be carefully selected to facilitate the reduction of pollutants. Normally, a mix of vegetation will optimise the uptake of nutrients. Consultation with plant material experts is essential for the success of wetland systems designed to remove pollutants. Field trials and laboratory testing of the vegetation will often be necessary prior to its introduction. This is particularly true if the vegetation is new to the region. Harvest considerations must also be made when selecting vegetation. The harvested biomass may have some value that will reduce the operational cost of the system.

Grass and reed types of vegetation are normally the most effective for improving water quality. However, some tree species may add diversity and habitat value to the wetland area if included in the site plans. The establishing of the proper vegetation type will probably require a temporary water management plan to facilitate early vegetative establishment. As vegetation selection guidelines have not been developed for every climatic region, or for each pollutant, field trials and tests are strongly recommended prior to constructing wetlands.

Pollutants generally move with the water flow, thus slow, continuous, uniform flow will provide the optimum pollutant uptake opportunity for the vegetation. The flow pattern should be designed to minimise evaporation in order to reduce the risk of developing water toxic to the ecosystem and to avoid channelling from inlet to outlet. The monitoring of water quality parameters is of utmost importance. For this to be done properly, it is important to establish the baseline situation. This means that an effort will be needed to determine the discharge quantity and quality of the water to be treated throughout the year prior to passage through the wetlands. Monitoring will be required to properly manage the wetland and to determine the effectiveness of individual wetlands.

Improved agriculture technology and irrigation water management will probably change the future composition and quantity of the discharged water. Thus, an interactive management programme should be developed that is effective for the conditions of the designed flow-through wetland. Institutional responsibility is important to the success of wetland systems for water quality and habitat protection. If the water quality control system is to remain effective, it will require numerous adjustments and proper maintenance.

It must be managed in accordance with the designs for the primary function of water quality improvement. However, secondary functions should not be ignored and often require some special management to achieve the benefits desired. The institution responsible for the wetland needs to have the financial means to operate, monitor and maintain the system in a sustainable manner. The one constituent that wetlands cannot remove is salt.

7

Drainage Water Re-use

Re-use is the natural method of managing drainage water. The major degradation factor of re-used waters is the high concentration of ions. Waters with low ionic concentrations provide plants with an adequate supply of many of the essential nutrients needed for growth. However, as salinity increases, specific ions may become toxic or interfere with the uptake of other nutrients. In soils, the accumulation of ions increases the osmotic potential against which plants extract water. It can also degrade soil structure.

Drainage and leaching of salts from the root zone are key factors in the management of salinity in agriculture. Another management factor is control of the range of salt tolerance expressed in crop species. Water re-use for agricultural crops has distinct economic incentives and a number of crops are known to be highly tolerant to salinity. However, as salinity increases in the irrigation water, there is a greater need to monitor and manage irrigation and drainage practices and to consider the sustainability of the system. It is possible to safely re-use agricultural drainage water if the characteristics of the water, soil, and the intended crop plants are known and can be economically managed.

Poor quality water requires selection of crops with appropriate salt tolerances, improvements in water management, and maintenance of soil structure and permeability (tilth, hydraulic conductivity). When sensitive crop growth stages such as germination and early growth are excluded, the temporal weighted mean root zone salinity has been found to be a valid measure for evaluating crop response to salt. The arithmetic mean root zone salinity within the rooting depth integrated over the time of exposure is an effective approximation for estimating crop response. Plants respond to the weighted mean salinity within a specific growth period.

Salinity

Salinity in water or soil is an environmental factor that, in general, reduces crop growth. At relatively low salinity, especially among crop species such as cotton or the halophytic sugar beet, some salinity may actually improve crop production. This effect has been attributed in some instances to an improvement in water use efficiency of the plant. However, as salinity increases beyond some threshold tolerance, yield decline is inevitable. Usually, at low to moderate salinities, plant growth is simply reduced and there is a slight darkening in leaf colour. These effects are difficult to detect without comparisons with non-saline controls.

When salt concentrations in the soil water reach toxic levels, leaves or shoots may exhibit visible symptoms of tip or edge burning or scorching due to high internal concentrations of salts. Other visible symptoms may be associated with nutrient imbalances caused by competitive interactions between Na^+ and Ca^{2+} or K^{4+}, or between Cl^- and nitrate. Interactions of salinity with nutrients such as PO_4^{3-}, Mg and micronutrients have also been reported, but these observations seem to be specific to certain crops and waters of specific ionic composition. Depending upon crop species and salinity concentration, salt in the crop root zone may also influence the rate of plant development by increasing or decreasing the time to crop maturity.

In most crops there are some notable differences in root/shoot ratio, a response that would not be identifiable in the field. In some crops, salinity changes plant growth habit or increases succulence. There is a wide range in plant species response to salinity. Sugar beet, sugar cane, dates, cotton and barley are among the most salt tolerant; whereas beans, carrots, onions, strawberries and almonds are considered sensitive. In general, salinity decreases both yield and quality in crops, and previous research has led to the development of large data bases on the salt tolerances of many crop species and varieties.

Salt tolerance can be represented most simply based on two parameters: the threshold salinity (t) which is expected to cause an initial significant reduction in the maximum expected yield (Y); and the slope (s) of the yield decline. Slope is simply the rate that yield is expected to be reduced by for each unit of added salinity beyond the threshold value. The formula to calculate relative yields is:

$$YR = Y - s\,(EC_e - t) \text{ where } EC_e > t \qquad (1)$$

Salinity is measured in units of electrical conductivity of a saturated soil paste extract (ECe) taken from the root zone of the plant as averaged over time and depth. Soil paste extracts are water taken from soil samples which are brought up to their water saturation points.

Electrical conductivities are measured on the filtered water extracts from these samples in units of deciSiemens per metre (dS/m). For most natural waters, 1 dS/m is equivalent to about 640 mg/litre salts. New methods use electronic probes or electromagnetic pulses to estimate ECe with less time and effort.

Where there is uncertainty in the determination of the t and s parameters in Equation 1, crops have been classified in a general range from sensitive to tolerant based upon available data. Crop salt-tolerance values and classifications are extremely useful in predicting the effects of re-using a water of a specific salinity on a particular crop. Most salt-tolerance data are based upon the effects of saline waters predominated by sodium chloride, sometimes with varied amounts of calcium as needed to avoid the development of soil permeability problems associated with soil sodicity. Drainage waters or waters re-used from agricultural processing or manufacturing operations may predominate in other chemical species and/or have high concentrations of B, Se, As or other ions that may be environmental hazards.

Management Methods

Traditionally, the goal of agriculture has been to maximise yield and profit, defined by the net difference between inputs and outputs. To maximise yields, soil salinity is reduced by leaching. The kind of management directed toward this goal is mediated by costs of water, drainage, applied nutrients and amendments to soils and waters. Where water quality and/or quantity is limited and where there are restrictions on drainage, the hazardous effects of salinity on crop growth and the dangers of either insufficient or excessive leaching are serious concerns. However, crops can be grown with saline waters provided that suitable irrigation and cropping strategies are used.

Management needs to be more intensive and more precise methods should be used for water application and distribution. Water requirements needed for crop use and leaching should be accurately assessed and provided in a timely manner. Most crop water use coefficients have been developed for non-saline situations. With saline water, growth and consumptive water use are reduced. However, preliminary evidence indicates that some crop water use coefficients may apply over a range of salinities because the water unused by the crop is needed to offset the increase in the leaching fraction requirement.

The greater the salinity of the irrigation water, the greater the need for adequate irrigation and drainage. Rates of salt accumulation in the soil are dependent upon the amount and concentration of the saline water applied and the amount remaining after plant water needs have been met. In a properly managed sustainable system, there may be a high salt content but no continuing accumulation. Soil salinity is spatially and temporally non-uniform. Salinity often rises during the summer and then decreases with winter rains. This is a desirable condition which helps reduce permanent salt accumulation. However, under such circumstances, correlating yield responses to such variable salinity is difficult.

Best management practices include land grading as a method to provide uniform surface irrigation, or the use of sprinkler or drip irrigation. Crops irrigated with sprinklers are subject to injury not only from salts in the soil but also from salts absorbed directly through wetted leaf surfaces. In general, plants with waxy leaves are less susceptible than others. When saline sprays come into direct contact with leaf surfaces, salts can accumulate to toxic levels in leaf tissue and cause decreased yields. In tree and vine crops, the extent to which leaves are wetted can be minimised by sprinkling under the canopy.

However, even with under-canopy sprinklers, severe damage to the lower leaves can occur. The extent of foliar injury depends on the concentration of salt in the leaves, weather conditions and water stress. For example, salt concentrations that cause severe leaf injury and necrosis after a day or two of hot, dry weather may not cause any symptoms when the weather remains cool and humid. As foliar injury is related more to frequency of sprinkling than to duration, infrequent heavy irrigations should be applied rather than frequent light irrigations. Slowly rotating sprinklers that allow drying between cycles should be avoided, as this approach increases the wetting-drying frequency.

Sprinkling should be done at night or in the early morning when evaporation is less. In general, poorer quality water can be used for surface-applied irrigation than for sprinkler irrigation. Drip irrigation gives the greatest advantages when saline water is used. Drip irrigation avoids wetting of the leaves with saline water and can be managed to maintain relatively high soil water potentials. As drip irrigation is normally applied frequently, there is a continuous leaching of the soil volume from which the plant extracts water. General leaching can be provided intermittently between growing seasons and it is supplemented by seasonal rainfall.

Leaching Requirement (LR)

In order to calculate the leaching requirement (LR), root zone salinity estimates are typically weighted to account for water uptake at different depths. The root zone may be conceptually divided by depth into four quarters, with a typical water uptake distribution pattern of 40:30:20:10 starting with the most shallow profile. This assumes a 'normal' rooting distribution but modifications should be made according to the frequency or type of irrigation (e.g., drip, sprinkler, alternate furrow). Of the various methods for calculating LR, the simplest is the equation proposed by Rhoades used in conjunction with salt-tolerance yield parameters:

$$LR = EC_i \ / \ (5EC\text{-}EC_i) \tag{2}$$

where EC_i is the electrical conductivity of the irrigation water, and EC is either EC_e at t or at the acceptable yield level below t, and LR is the fraction of the irrigation water that must be passed through the root zone to control salinity.

Other Management Methods

There are a number of cultivation and management methods which may offset yield loss when saline water is used for crop production. Some simple methods include: using more vigorous cultivars; using screens to select seed of larger size and higher seed weight; increasing population densities to offset smaller plant size and reductions in tiller number; and placing seed on sloped or modified seed-beds so that salts drawn to the surface by evaporation accumulate away from the seed line. Another way to move salts away from the seed or plant base is to irrigate in alternate furrows.

An additional method of ameliorating the effects of moderately saline drainage water is to match fertilisation requirements with the chemical composition of the water and soils. As drainage waters often contain some of the essential fertilisation requirements of plants, applied fertilisers can be reduced. On the other hand, as ion interaction and competition reactions affect plant uptake, additions of calcium, potassium or some other element may be required for better crop growth. An important consideration is that high sodium concentrations in soils and waters may decrease soil permeability and prevent effective leaching of salts.

The suitability of soils and waters is measured using a parameter called the sodium adsorption ratio (SAR). Soil amendments are sometimes a necessary part of salinity management schemes to reduce SAR. On the other hand, high EC (salt concentrations) of the water increase the soil permeability.

Cycling and Blending Strategies

Cyclic strategies for using waters of different salinities have been proposed and it has been demonstrated that cyclic irrigations can be successfully applied to crops during different growth stages or can be used with crop rotations between tolerant and sensitive crops. The feasibility of cyclic and blended applications of high quality water with drainage water depends on both supply and the availability of storage, mixing and delivery systems. Where non-saline waters are available for critical irrigations, growers can take advantage of the fact that many crops are most salt sensitive during the germination and seedling stages and are much more tolerant during later growth stages.

Plant salt tolerance varies with phenological growth stage, and beneficial effects can be obtained by managing irrigations to take advantage of this fact. For example, at salt-sensitive growth stages, salinity effects can be reduced by providing higher quality water or by minimising salt accumulation through increasing the number of irrigation cycles. In some crops, higher salinities applied during the later stages of development will improve yield quality by increasing the content of sugars or soluble salts in the fruit.

The timely application of saline water during fruit development has been used as a strategy to improve both sugar content in melons and soluble solid content in tomatoes. In response to some moderate levels of salt and drought stress, fruit trees sometimes exhibit significant increases in fruit set and yield, but usually at a cost in subsequent years due to reduced biomass production. Another disadvantage may be decreases in shipping quality. Francois and dark found that increasing salt stress delayed fruit maturation in 'Valencia' oranges but did not effect fruit quality. Their results also indicated that salinity had no effect on total soluble salts (TSS), but Bielorai et al. measured slight increases in TSS and sugar contents of 'Shamouti' oranges when the chloride concentration in the irrigation water was 450 mg/litre (m g/g).

For cyclic use strategies, factors that should be considered include the effects of changes in salinity during the growing season, the average salinity distribution in the root zone, the interactions with climatic variables, and the effects of different soil types. After crop establishment, the salinity in the root zone averaged over time may be considered as the effective salinity exposure. This means that in most cases, once crops are established, there will probably be little measurable difference in a field situation as a result of applying 4 and 10 dS/m water in alternate irrigations applied over a crop cycle as opposed to using a water of 7 dS/m in each irrigation. This is the result of compensating factors in both soils and plants. Models are being developed to predict the effects of different irrigation water salinities during the growing season on crop yield.

Blending is the mixing of poor quality drainage water with good quality irrigation water. Provided that the blended water is sufficiently low in total salinity and toxic ions, this is the most economic and environmentally acceptable means of disposing of drainage water. This strategy can potentially increase the salinity of the groundwater over time. Often, separation of waters of different qualities is the best strategy. Egypt has an extensive drainage water re-use programme. Over 4, 000 million m^3 of agricultural drainage water, produced in the upper Nile Delta, are re-used to supplement irrigation water requirements in the lower delta. Pumping stations lift the better quality drainage water into irrigation canals, where mixing occurs. The Ministry of Public Works and Water Resources has an extensive re-use quantity and quality monitoring programme.

Agriculture-forestry Systems and Solar Evaporators

The agriculture-forestry system and solar evaporator are still at the development stage. The concept compares favourably with other intermediate disposal technologies. It manages drainage water as a resource instead of directly disposing of it into evaporation ponds, rivers or the ocean. Agriculture-forestry systems produce energy (biomass) while some other technologies for the treatment of drainage water have high energy demands. By significantly reducing the final volume of non-re-usable drainage water, agriculture-forestry systems provide an opportunity for seasonal discharging of drainage water into rivers during high water flows.

Solar evaporators are preferable from the standpoint of wildlife safety as they provide for better bird control than do evaporation ponds. The trees create wildlife habitats, reduce air pollution, and enhance the overall environmental quality of fanning regions. The agriculture-forestry method has been developed and tested in California for salt management on irrigated farmland. It has two objectives:

- to utilise the drainage water as a resource to produce marketable crops;
- to reduce the volume of discharged drainage water directly on the farm.

The agriculture-forestry system for the productive use of drainage water and its disposal into a solar evaporator has been developed in semi-arid conditions in California and other regions. A diagram of the system is shown in Figure 1. The main function of the trees is to use and evaporate large volumes of drainage water. This can be achieved not only through sequential re-use but also by uptake of the drainage water from shallow water tables, or by intercepting the flow of drainage water from upslope.

The trees can be viewed as biological pumps. The tree biomass offers a number of marketing options ranging from electricity generation to the production of biofuels or biochemicals. Drainage water is further concentrated through the irrigation of halophytes that also have marketing potential as food or industrial crops. The salt concentration in the drainage water in each succeeding stage is increased, but the volume of water is reduced. About 80-85% of the initial volume of drainage water produced while growing salt-sensitive crops is sequentially re-used to produce salt-tolerant crops. The

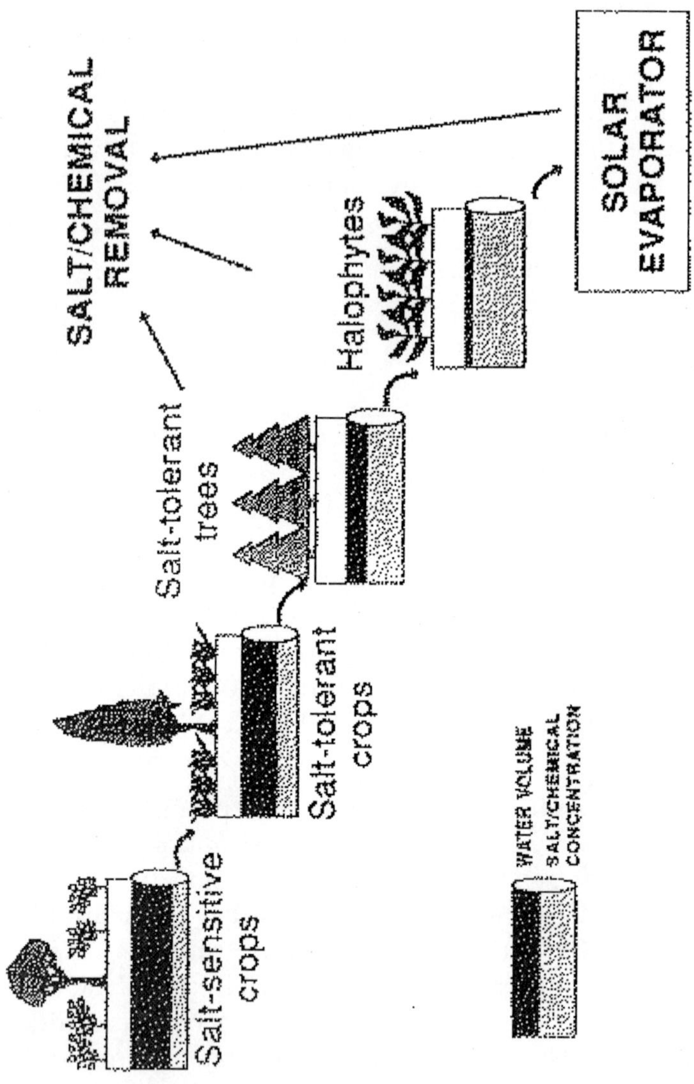

Figure 1. Agriculture-forestry and salt management systems

remaining 15-20% of drainage water, with increased concentration of salt, evaporates in solar evaporators.

The design of the solar evaporator consists of a levelled area lined with a plastic liner on which the crystallised salt is collected The daily amount of drainage water discharged into the solar evaporator is correlated with daily evaporation rates to prevent water ponding. This makes the facility unattractive to wildlife. This is important as the concentration of Se has proven to be a hazard to wildlife. The agriculture-forestry system for the sequential re-use and reduction of drainage water volume can be viewed as the flow of water from salt-sensitive crops to salt-tolerant trees to more salt-tolerant halophytes and to a solar evaporator (Figure 1).

The water flow is programmed and automated using a system of sensor-controlled equipment. The agriculture-forestry system operating in California for sequential re-use of water and salt removal utilises drainage water with an initial salt concentration of about 7 000 mg/litre, following the irrigation of salt-tolerant crops. This concentration of drainage water has been conventionally disposed of into evaporation ponds, rivers and/or the ocean. The agriculture-forestry system concentrates salt in a significantly reduced volume of drainage water. This technology offers management options of salt crystallisation in a relatively small area on farms (solar evaporators) or the discharge of a reduced volume of drainage water (brine) into solar ponds or natural sinks (e.g., the ocean).

The crystallisation of salt in solar evaporators provides management options for salt marketing or its (short/long-term) storage on farms or designated landfills for storage in perpetuity. The system for sequential re-use of drainage water is also effectively reducing the level of

Se. A typical Se concentration is 0.5 mg/litre in drainage water applied to trees, and 0.9 mg/litre in the reduced volume of drainage water applied to halophytes. Consequently, the estimated quantity of Se is about 50 kg per 100 000 m^3 of drainage water applied to trees, and it is reduced to about 27 kg per 30 dam^3 of drainage water sequentially re-used to irrigate halophytes.

Selenium is reduced through volatilisation and is also taken up by trees and halophytes. Selenium removal by trees is mainly concentrated in leaves, and the measured values (in terms of mg/kg) were 0.5-0.9 for eucalyptus, 0.6-1.0 for casuarina, 2.6-3.6 for athel, and 0.6-1.8 for halophytes. Successful experiments were conducted on the transfer of Se through harvesting halophytes and safely using this forage for cattle feeding.

The design of a solar evaporator is fundamental to the development of an integrated agriculture-forestry system. The size of the solar evaporator is a function of evaporation rates typical for the farming region. The area of the solar evaporator is calculated by:

$$Ae = \frac{1{,}000 DW_h}{EV \times C_e}$$

where,

Ae area of the solar evaporator (m^2)

DW_h drainage water discharged from halophytes into the solar evaporator (thousand m^3)

EV annual evaporation rate (m)

C_e coefficient of evaporation, for reduced evaporation of saline water.

Further research data are needed to estimate the coefficient of evaporation (C_e). Experimental data indicate that the salt concentration in the drainage water discharged from halophytes into a solar evaporator ranges from 32, 000 to 41, 000 mg/litre. The final concentration of the crystallised salt exceeds 180 000 mg/litre. The major components of crystallised salt are SO_4^{2-} (56.70 %), Na (23.50 %), and Cl^- (8.40 %). The Se content ranges from 2 to 8 mg/kg. The area of halophytes that can be grown is a function of the quantity and quality of drainage water recovered from the trees.

Experimental data collected during a period of six years indicate that the increase of salt concentration in water drained from trees is about 2.8 times higher than of the drainage water (from conventional farm crops) applied to trees. This indicates that approximately 65% of the re-used drainage water is consumed by trees. The applied 'tree' water (drained from farm crops) is typically EC 8-10 dS/m and the applied 'halophyte' water (drained from trees) is about 28-30 dS/m.

To maintain adequate control of soil quality, the leaching fraction (the percentage of the infiltrated irrigation water that percolates below the root zone) must be sufficiently high to prevent a build-up of salts, Se and B in the soil profile. The SAR should also be monitored. An agriculture-forestry area of 16 ha has the capacity to process about 110 000 m^3 of drainage water per year. The size of a farm that can be serviced by the agriculture-forestry system (including solar evaporator) is a function of several factors, such as cropping system, quality of irrigation water, irrigation water management, soil salinity, and the use of trees for controlling groundwater conditions (water uptake from high water tables or subsurface flows).

It is estimated that 30 ha of trees, halophytes and solar evaporator can utilise and process drainage water from about 1 000 ha of conventional farm crops. The 30 ha area would consist of 18.75 ha of trees, 7.5 ha of halophytes, and 3.75 ha of solar evaporator. The relative size of areas is as follows: irrigated area, 1 000 ha; trees, 10 ha; halophytes, 4 ha; and solar evaporator, 2 ha. The management of biomass from the trees and halophytes is essential for the performance of agriculture-forestry systems and salt removal.

The trees and halophytes need to be selected for their salt tolerance, which should range from about 9 000 to 18 000 mg/litre for trees and 18 000 to 41 000 mg/l for halophytes. The halophytes should preferably be perennial plants. The other required characteristics of trees and halophytes include: high water demands, tolerance to frequent flooding, frost tolerance, and marketability of harvested biomass. Eucalyptus camaldulensis is the species of choice for this particular salt management system because of its salt tolerance and high water requirements.

To improve the quality of eucalyptus trees for agriculture-forestry sites in the San Joaquin Valley, a selective breeding programme was initiated in 1987. Selected trees have been systematically evaluated each year, and 22 trees have been chosen for tissue culture propagation. The programme is seeking to achieve a higher diversification of salt-tolerant trees. Additional experimental trees planted include casuarina, athel, acacia, mesquite and poplar. The selection of halophytes has been based on literature review, field evaluation trials, and a survey of salt-tolerant plants in California. These plants are being selected not only for salt management purposes, but also with a consideration for their biological interaction with conventional farm crops. This is to avoid introducing species that could be potential weeds or host plants for insect vectors or plant viruses.

Halophytes have been selected for salt tolerances ranging from an EC of 20 to 45 dS/m. Based on current field evaluations, the most promising plants include salicornia, iodine bush (Allenrolfea occidentalis), salt grass (Dichtilis spicata), and cordgrass (Spartina gracilis). Other promising halophytes include fivehook bassia (Bassia hyssopifolia), Jose tall wheatgrass (Agropyron elongatum), fat-hen (Atriplex patula), red sage (Kochia americana), Atriplex nummularia and Atriplex lentiformis.

Re-use in Wetland Habitat

Drainage water can also be used for wildlife or wetland habitat irrigation. However, prior to using agricultural drainage water for this purpose, the following questions should be addressed:

- — Is the water from a surface or subsurface drain system or both?
- — What types of vegetation are to be grown?
- — What constituents are in the water?
- — Is the water available when it is needed?
- — Is an adequate volume of water available?
- — Will it have positive impacts on wildlife and the environment?
- — Will there be adequate runoff from the wetland?
- — Is the wetland sustainable?

Re-use of Surface Drainage Water

Where drainage water is derived from only surface

drainage or tailwater sources, the main question is whether or not the water contains applied and persistent pesticides. In areas where strong environmental safeguards exist and pesticide container label restrictions are followed, there is little risk associated with the re-use of surface runoff or tailwater drainage water. Most of the surface derived drainage water is used to flood wetlands in the early autumn. In the rice growing regions, the fields are drained in the late summer or early autumn. This drainage pattern coincides with the autumn migration of the waterfowl to their wintering grounds.

Ideally, the winter waterfowl habitat is flooded to a depth of 20-50 cm. Depending on soil type, seepage and evaporation rates, the drainage water required for initial flooding will range from 500 to 1,500 m^3/ha. Where local supplies of surface derived drainage water are available, water is used to maintain ponds from October to March. In warm climates, the annual evaporation or consumptive use is approximately 2,500 m^3/ha, which causes a further salt concentration increase in the wetland outflow. Typical native marsh plants grown with surface agricultural drainage waters include: smartweed (Polygonum lapothefolium), swamp timothy (Heleochloa schenoides), tule or hardstem bulrush (Scirpus fluviatilis), and cattails (Typha spp). These plants are grown under a moist soil water management regimen.

Water is applied in the autumn of the year and held until spring soil temperatures begin to warm. This occurs in March or April in the Central Valley of California. When the soil begins to warm, the ponds are drained to mudflat conditions. This stimulates seed germination. In some areas they require no additional water until the autumn flood-up period. However, where summers are hot and dry, they will require one or more irrigations in

July or early August to provide for optimal seed production for migrating wildlife food.

Re-use of Subsurface Drainage Water

The re-use of subsurface saline agricultural drainage water for wetland management poses substantial challenges and can generate problems which could result in wildlife losses and habitat reductions. Although subsurface saline drainage water is typically free from contamination by applied pesticides or herbicides, it may contain soil or naturally derived toxicants or trace elements such as salts, nitrates, As, B, Cd, Cr, Pb, Hg, Mo, Ni, Se, Ag, U and V. Each of these constituents is potentially toxic independently, in combination with other constituents, or through the process of biomagnification in wildlife through the food chain.

Careful analysis of subsurface agricultural drainage water during several periods of a yearly cycle is required before any plans for re-use as a water supply for wetlands can be considered. The provision of an adequate volume of flow-through water is important to minimise concentration of toxic elements due to evaporation. At present, there are no comprehensive standards which establish safe levels of trace elements in water used for wetland habitat management. However, because of the high potential for food-chain magnification, most wetland managers intentionally refuse to use subsurface agricultural drainage waters which contain levels of trace elements above background levels.

The potential for, and the costs of, clean up or remediation of a contaminated wetland dictate a conservative approach. Where trace element contamination is not a concern, saline drainage water can be used to support a productive wetland habitat. The major consideration is the management of soil and water salinity. The maintaining of a salt balance between the applied water and the soil/water interface is key to the production of brackish water native marsh plants.

In general, water with a TDS level of 2,500 mg/litre or less is preferable for wetland management. Sometimes, water with a TDS level as high as 5,000 mg/litre can be used for short periods. Standard management practices involve an autumn flooding to a depth of about 20-50 cm, with these depths being maintained until January or February. In late winter, the ponds are drained to discharge drainage water and accumulated salts. The ponds are then refilled with new water as deeply as practicable. After approximately 14 days, the water is drained again.

The drainage cycle is repeated two to three times before the cycle is completed. This process removes salt concentrated in the surface water through evaporation and allows for a rebalancing of the water/soil salt equilibrium. The ponds are drawn down to a mudflat state in March or April to facilitate germination of desirable salt-tolerant marsh plants. Typical plants grown under this regimen include: alkali bulrush (Scirpus robustus), brass buttons (Cotula corinopifola), salt grass (Distichilis spicata) and tules {Scirpus acutus). Depending on the local soil and climate, one or more shallow irrigations may be necessary to bring germinated plants to full maturity. In all cases, in order to prevent excess salt accumulation, water circulation is maintained when the ponds are filled with saline drainage water, and some constant rate of discharge or outflow from the ponds is necessary.

By maintaining water circulation and some constant rate of water flow, outbreaks of waterfowl disease can also be avoided. Any wetland habitat supported principally with saline agricultural drainage water must be carefully managed and monitored to have productive wetlands. In addition, there must be an environmentally safe way (ocean, large river, salt lake) of disposing of the water as it is drained from any wetland area.

8

Drainage Water Issues

Careful control and appropriate reuse of drainage water can help protect the environment and optimise the use of water resources. The health issues related to drainage water management can be grouped in three categories:

— water related vector-borne diseases;

— faecal/orally transmitted diseases; and

— chronic health issues related to exposure to residues of agrochemicals.

In tropical and subtropical regions there is a close link between the presence of excess water (due to lack of adequate drainage) and the transmission of water related vector-borne diseases. Malaria, schistosomiasis (bilharziasis) and lymphatic filariasis are important water related vector-borne diseases. Despite control programmes, health services and available treatments, these diseases today represent a growing health problem. Water related vector-borne diseases are caused by bacteria, viruses and parasites (protozoa and helminths) transmitted by water related disease transmitting agents, also called vectors or intermediate hosts. A vector is an animal, often an insect, that transmits an infection from one person to another person or from infected animals to humans.

Most infections can only be transmitted by a particular, disease-specific vector, e.g., malaria by Anopheles mosquitoes. An intermediate host has a similar role to a vector. However, such an organism does not actively transmit a pathogen, like freshwater snails in the case of schistosomiasis. Vectors and intermediate hosts represent critical elements in various disease transmission cycles of parasitic water related diseases. In general, they live in or near aquatic environments. Direct pathogen transfer and the transmission by vectors and intermediate hosts require specific environmental and socio-economic conditions. The conditions are defined by:

- — quality and quantity of water;
- — type and frequency of human-water contacts;
- — number and distribution of vector or intermediate host breeding sites; and
- — exposure of humans to vector and intermediate host populations.

Consequently, the above-mentioned diseases can also be associated either directly or indirectly with the design and management of treatment and disposal plants for the re-use, treatment or disposal of drainage water. The key criteria for such a health risk are:

- — introduction of temporary or permanent open water surfaces bodies, e.g., constructed wetland, stabilisation ponds, or evaporation ponds;
- — suitability of such water for vector breeding;
- — accessibility for the local population;
- — location in relation to human settlements and transport links (e.g., roads); and
- — pollution by organic or inorganic substances.

Misuse and lack of maintenance are the two main reasons why drainage structures (road drainage ditches, culverts, dam site drainage or drainage canals in irrigation schemes, and also drainage water treatment and disposal facilities) are often associated with environmental health problems. Farmers, associations or national agencies generally conduct regular maintenance on irrigation canals. Water quality and flow velocity are relatively high. However, in drainage facilities the opposite conditions are frequent. Silting, uncontrolled aquatic weed growth, slow water flow or stagnant pools associated with the resulting wetlands offer ideal breeding conditions for mosquitoes and aquatic snails.

Farmers seem to concentrate on irrigation water management rather than on drainage management. Moreover, there is often a lack of adequate domestic water supplies and sanitation facilities. Thus, drainage canals or drainage water treatment and disposal facilities are often misused for washing, drinking and uncontrolled disposal of human excreta or other waste by the poorest and, thus, most vulnerable social groups. In this way, drainage water contributes to disease transmission.

Water Related Diseases

Although the eradication of diseases such as malaria has long been an objective, the problem is far from being solved. According to the most recent information the malaria situation is serious and deteriorating. Global malaria mortality is estimated at 1.5-2.7 million and global malaria cases at 300-500 million. Malaria is one of the most serious health problems facing African countries and a major limitation on their socio-economic progress. Children under the age of five and pregnant women are most at risk. WHO reports that 90 percent of the global burden of this disease can be attributed to environmental factors, including land and water management. Schistosomiasis is almost as widespread as malaria but rarely causes immediate death.

An estimated 200 million people are infected and transmission occurs in about 74 countries. The infection is primarily common in children who play in water inhabited by the snail intermediate host. Water development projects, especially those associated with the irrigation of large areas, have often been associated with an increased incidence of schistosomiasis. Intestinal schistosomiasis was unknown or infrequent in the Nile, Senegal and Volta deltas before the construction of the Aswan, Diama and Alosombo dams. Worldwide parasitic and infectious communicable diseases (including all water related diseases) caused 32% of all deaths in 1993, resulting in the loss of 16.5 million lives; 99% of reported cases occurred in developing countries. In these countries, parasitic and infectious communicable diseases cause 41.5% of all deaths.

Vector-borne Diseases

Insect vectors represent the largest group of disease transmitting agents. In most cases and for the most widespread diseases, mosquitoes are the main vectors. Among a wide range of vector-borne diseases, two diseases, namely, malaria and lymphatic filariasis stand out as serious health hazards in the context of poor drainage.

Malaria

Malaria is caused by a protozoan parasite of the genus Plasmodium. Malaria is a complex disease causing fever, anaemia and an enlargement of the spleen. This causes additional cerebral complications, especially for children. Correspondingly, child mortality rates for P. falciparum, one of the four types of malaria affecting humans, are very high with approximately one million children below the age of five dying in 1993. Malaria is transmitted by the bite of a mosquito of the genus Anopheles. The transmission cycle is only between man and mosquitoes.

Man acts as the intermediate host or reservoir and the mosquitoes as the vector. Protozoan parasites of the genus Plasmodium have to undergo complex development and multiplication processes both in man and mosquito before they can be further transmitted. Only the female mosquitoes are of importance for transmission, as they need a blood meal for oviposition. Malaria covers not only all developing countries, but is present on almost the entire land surface between the latitudes 40°N and 60°S. However, the distribution is not uniform and depends mainly on climate, altitude, population density and the specific environmental requirements of the mosquitoes species.

Highly endemic areas are sub-Saharan Africa, Central America and the northern part of South America, the Indian subcontinent and Southeast Asia. In the last two decades, a growing number of malaria cases have been observed. However, this cannot only be explained by the increasing population. To a large extent, this is also due to the increasing number of WRDPs such as irrigation and drainage schemes or hydro-electric dams. With the introduction of new open water surfaces in the form of canals, ponds and artificial lakes, new mosquito breeding sites have been created. However, the persistence of the disease is also due to the absence of effective long-lasting vaccines, and the growing resistance of malaria pathogens and mosquitoes to treatment and insecticides, respectively.

By the end of 1985, 50 of the 150 potential malaria transmitting Anopheles species were already recorded to be resistant to one or more pesticides (including DDT). At least 11 of those 50 species are known to be important and dangerous malaria transmitters. Furthermore, malaria transmission is not only related to WRDPs. Deforestation, mining, road construction and all the negative consequences of rapid and uncontrolled urbanisation are also contributing to the creation of mosquito breeding sites. In this context, urban drainage plays two key roles.

On the one hand, it is an essential and effective tool for reducing and eliminating mosquito breeding sites by controlling surface water and waterlogging and by eliminating unnecessary open water surfaces. However, on the other hand poorly maintained drainage canals can represent potential breeding sites for various mosquito species if they are permanently flooded and aquatic weeds are not cleared. Finally, a fresh risk might result from climatic change. First reports suggest that global warming can change the geographical distribution of mosquito breeding and shift the malaria transmission border line to the north.

Lymphatic filariasis

Mosquito-borne filariasis includes a group of diseases which lead to inflammations and obstructive lesions of the lymphatic system. Filarial parasites are nematode worms, which also need an insect to complete their life cycle. The main disease is Wuchereria bancrofti or elephantiasis and is transmitted by mosquitoes of the genera Culex. During a bite, a mosquito might take up a number of microfilariae which then undergo a development cycle in the mosquito. When the infected mosquito bites again, the infectious larvae enter first the skin and then the lymph vessels and lymph glands.

Unlike malaria, filarial infection requires repeated and lengthy exposure to the vector due to the low pathogen load per bite. For W. bancrofti, long exposure to repeated infection will finally lead to severe and disfiguring deformities of the legs, arms and genitals. Lymphatic filariasis affected about 120 million people in 1994; 80% of cases were reported either from tropical Africa or India. Although filariasis is more of a serious urban health problem, it can cause significant health hazard in rural settlements and farming communities where poor surface drainage, lack of sanitation facilities and environmental pollution through uncontrolled waste and excreta disposal are prevalent. The principal vector, Culex quinquefasciatus (fatigans), breeds in polluted water in ditches, drains, tanks, barrels, tins and any kind of water accumulation container.

Mosquito bionomics

There are more than 3,000 known mosquito species. However, of the 150 species that are potential vectors, only 30 are considered dangerous. The three genera Anopheles, Aedes and Culex, from three sub-families, are disease relevant. The epidemiology and the life cycle stages demonstrate the importance of climatic factors and that water is the essential environmental component for mosquitoes. The immature stages such as eggs, larvae and pupae require an aquatic environment, whereas adult mosquitoes live in terrestrial ecosystems.

The quality and quantity of water, whether it is running or standing, shallow or deep, clean or polluted, sweet or brackish, shaded or sunlit, permanent or seasonal, and finally the climate will determine which particular species can breed. Some of the physical and biological factors are given below.

Physical factors: In general mosquitoes prefer higher air humidity and average water temperatures between 23°C and 33°C in order to complete their aquatic stages within two weeks. Rainfall can be a limiting as well as a positive factor. Rain will fill ditches, rivers, ponds, etc., but heavy rain can have a flooding effect and flush out breeding places. Mosquitoes do not normally reproduce where excess water is quickly removed. Sunlight or shade can also be positive or negative depending on the species. All these factors also determine the resting and biting habits. Some species rest and feed indoors and/or outdoors. Peak biting activity is usually about one hour before dawn. However, Aedes is a daytime biter, and many Anopheles bite throughout the night.

Biological factors: The presence of vegetation and floating plants are important for optimal breeding conditions. First, the plants are larval food and, more importantly, they provide shelter from predators and protection against wave movement. Therefore, mosquito larvae are not found on the open surfaces of large water bodies. The abundance of a number of species is linked to the presence of specific plants.

As a group, mosquitoes breed in an almost infinite variety of sizes, types and qualities of water bodies and each species requires specific environmental breeding and living conditions. However, most of the mosquito vectors breed in a rather restricted and narrow range of habitats. It is the number of potential transmitting species and their population dynamics which makes control efforts difficult. A comprehensive classification by species, country and habitat, including potential environmental management measures, is presented in the Manual on Environmental Management for Mosquito Control. Natural biological control of mosquitoes can be accomplished with larvivorous fish.

Over 250 different species of such fish are known to consume mosquito larvae. The most common mosquito fish is Gambusia affinis which is found in about 60 countries. Mosquito fish can be propagated commercially and introduced into drainage water. The fish are most effective during the warmer months of the year.

Schistosomiasis

There are three major types of schistosomiasis that affect man: S. haematobium, S. mansoni and 5. japonicum. The disease is caused by female and male trematode worms inhabiting the blood vessels of the urinary bladder (S. haematobium, S. japonicum) or the portal and mesenteric veins (S. mansoni). Typical visible symptoms of schistosomiasis infection are blood in the urine in the case of S. haematobium, and intermittent diarrhoea and faeces containing blood in the case of S. mansoni. Although effective chemotherapy is available, the costs per treatment and per caput make it too expensive for many developing countries. The transmission of all three species is based on a complex four-phase cycle which includes the presence of freshwater snails (Figure 1).

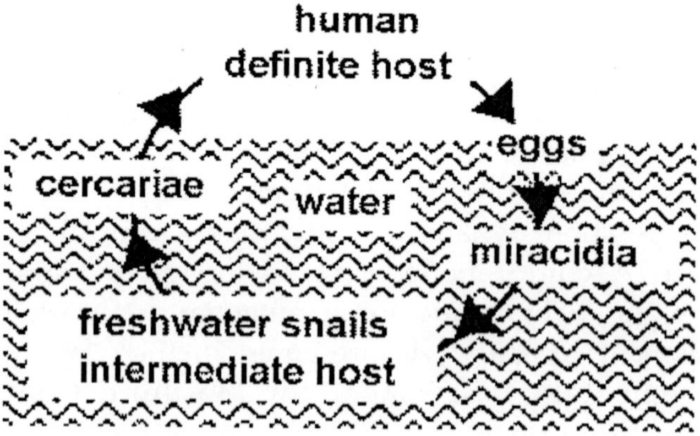

Figure 1. Transmission cycle of schistosomiasis

Eggs expelled in the urine or faeces of an infected person may reach water, where they hatch rapidly into miracidia, a free-swimming brief larval stage. Depending on the species, the miracidia have to find a specific aquatic or semi-aquatic freshwater snail as intermediate host within 24 hours. S. haematobium, S. mansoni and S. japonicum require snails of the genera Bulinus, Biomphalaria and Oncomelania, respectively. Oncomelania are semi-aquatic snails that live part of the time outside water in moist soil or mud. In the snails, the parasites develop within 4 to 6 weeks into cercariae, a second free-swimming larval stage. It is at this stage that they are infectious for man.

Snails shed numerous cercariae into the water. These have to find human beings in contact with water in order to penetrate their skin. Having penetrated the skin of the host, the parasites find their way through the veins, heart and lungs to the final organ, there recommencing the cycle with egg production.

Bionomics of snail intermediate hosts

Although individual snail species require a specific physical environment, the variety of aquatic habitats is almost infinite. Marshes and swamps, permanent or temporary ponds and pools, natural or man-made freshwater lakes or reservoirs, seasonal or permanent or slow flowing river streams, irrigation or drainage canals, rice fields and all other types of standing, slow flowing or impounded water are potential snail breeding sites. Generally, the water needs to be shallow, clean or brackish, with little turbidity. The duration of the life cycle depends on water temperatures.

The ideal water temperature ranges from 26° C to 28 °C. Snail intermediate hosts of both S. haematobium and

S. mansoni show great tolerance regarding pH values (5.3-9), mean water temperatures (18°-30°C) and salinity. Snails are found in shaded water bodies but are also known to be active when exposed to direct sunlight. The shedding of cercariae correlates with day-time and thus the intensity of sunlight. Furthermore, snails have a remarkable capability to survive long periods (5-8 months) in moist sand or mud (aestivation).

Their main source of food is organic matter originating from decaying submerged or emerged vegetation, different species of algae, bacteria and fungi. However, both of these aquatic snail species are sensitive to water velocity and water table fluctuations. Tolerable average current speed ranges between 0.0-0.3 m/s. In natural rivers, snails are generally dislodged due to turbulence and fleeting shear stresses along the transition zone between the embankment vegetation and the river bed sediments, which are their actual habitat zone.

Water-borne excreta related infections

Most of the aquatic vector or intermediate host habitats correspond to sites where humans access water for washing, bathing, swimming (mainly children) or fishing. Cairncross and Feachem proposed a classification of water-borne diseases in order to more fully understand the effects of excreta disposal:

— *Faecal-oral diseases (non-bacterial)*: Infections transmitted from person to person due to domestic contamination and lack of personal hygiene. Examples: hepatitis A, giardiasis.

— *Faecal-oral (bacterial)*: Person-to-person transmission with longer transmission cycles either through contaminated food, crops or water contaminated with faecal material. Examples: Various diarrhoeas and dysenteries like cholera, E. coli diarrhoea or salmonellosis, enteric fevers such as typhoid.

— *Soil transmitted helminths*: Eggs of parasitic worms are expelled in faeces and require a development stage in moist soils. They reach the human host either by being ingested on vegetables or by penetrating the soles of the feet. Transmission takes place in communal defecation areas or around dirty latrines without clean concrete floors. Examples: ascariasis, trichuriasis, hookworm.

— *Beef and pork tapeworms*: Transmission cycle includes an intermediate development stage in an animal and infection of man occurs when the meat is eaten without sufficient cooking. Transmission can be triggered through the application of sewage sludge as fertiliser on grazing land. Examples: taeniasis.

— *Water-based helminths*: Eggs in faeces must reach water in order to undergo the next development stage in an aquatic snail. The guinea worm (fasciolopsiasis) follows a similar cycle.

— *Excreta related insect vectors*: Filariasis transmitting Culex mosquitoes prefer to breed in highly polluted water. Badly maintained, unventilated latrines or uncovered septic tanks offer the best breeding conditions, mainly in urban areas. However, filariasis may also reach man simply by being carried by flies or cockroaches.

An important characteristic is the persistence of a specific pathogen, i.e., its ability to survive in the environment and whether animals, either in series or parallel, are part of the transmission cycle in view of the above classification, water quality and the environmental conditions around water bodies can be seen to play an essential role in the transmission. Thus, unlined open drains, stabilisation and evaporation ponds or wetlands might become particularly dangerous transmission sites. It is at such sites that excreta disposal, animal feeding and human water contacts might occur in a concentrated and uncontrolled form, so leading to the necessary pathogen-man/animal contacts.

Chemical Pollution and Health Problems

Health risks related to water quality, primarily due to agrochemicals, represent a growing problem. This is mainly due to the intensified production and the expansion of irrigated agriculture, often in association with food processing industries. Both subsurface and open drains can have an impact on human health on four levels:

— Drainage water re-use. Leaching or surface runoff from fields treated with pesticides, e.g., herbicides, fungicides, insecticides or molluscicides, represents a non-point source of toxic organic substances. If re-used as irrigation water, toxic substances can be ingested by individuals directly or indirectly in the case of drainage water being mixed with drinking water resources. Another source of direct water contamination with agrochemicals is the washing of spraying equipment in open drains or irrigation canals or the spraying of irrigation canals and reservoirs against aquatic weeds. Leaching also includes the transport of inorganic substances like salts, nitrate, phosphorus or heavy metals. Such substances can lead to health problems if consumed in high enough doses. However, salt will be less of a health problem as a salty taste will prohibit individuals from drinking too much salt polluted water. Trace elements are more of a concern in the case of subsurface drainage water.

— Drainage water treatment and disposal. Both lead to an accumulation, a transfer or a transition of pollutants. In addition, treatment facilities such as constructed wetlands or stabilisation ponds introduce new and permanent open water surfaces, and aquatic weed growth. Here, the question is whether this will change the human-water contact patterns and create new vector breeding sites. Furthermore, the quality of inadequately treated drainage water might also represent a potential health risk.

— The application of insecticides or molluscicides during chemical vector control campaigns. This can lead to short-term high concentrations of toxic chemicals in irrigation or drainage canals. Unprotected workers are exposed to health risks and local populations might consume polluted water without being aware of the risks. Indirect contamination might occur through the ingestion of fish.

— Uncontrolled disposal of industrial liquid waste in drainage canals or treatment and disposal systems. Due to the lack of environmental regulations and proper treatment plants, especially in urban areas, drains often serve as liquid waste disposal sites for local industries, thus representing dangerous point sources of toxic substances.

Although health risks arising from chemical pollution are accepted as a major environmental concern, data on medical implications are difficult to obtain and often unavailable for developing countries. Due to the lack of environmental protection and regulation standards as well as the lack of environmental monitoring data, governmental services often neglect the issue of pollution control.

Control of Water Related Diseases

Control of water related diseases always requires the interruption of transmission cycles, which include a number of quite different actors: man, animals, or vectors. These all have different functions, react differently to environmental changes and have different capacities in either transmitting a disease or in resisting infection. There are basically three ways to disrupt transmission:

— by eliminating or reducing vector densities;

— by protecting the susceptible host through immunisation, prophylactic drugs, reduced man-vector contacts using repellents, window screening, bed nets, or by installing adequate water supplies and sanitation facilities; and

— by reducing the reservoir of infection, by treating infected people or animals and by eliminating the pathogen in the host body.

Therefore, an effective and integrated disease and vector control strategy has to integrate four basic elements: chemical vector control, biological vector control, environmental management, and medical treatment. Medical treatment is still considered as most important. In fact most research and financial resources are focused on medical or bio-medical issues. However, it is clear from the magnitude of the present health problems, the increasing number and spread of diseases and the close environmental interactions set forth above, that any intervention can only be successful when all four elements are integrated. This requires all parties involved to have a sound understanding of all environmental, social, economic and bio-medical implications of the epidemiological dynamics of water related disease transmission.

The core of such an approach can be reduced to the formula: transmission and morbidity control. Transmission control combines chemical, biological and environmental management control of vectors, intermediate hosts, or any activity in the field of water supply, sanitation facilities and water treatment. Morbidity control includes chemotherapy or any medical control of the pathogen and the treatment of the disease. Oomen et al. visualise the environmental management component as a vehicle having hygiene education and community participation as wheels rolling on the road of basic infrastructure.

In 1979, WHO, FAO and UNEP established the Panel of Experts for Environmental Management (PEEM). The aim was to better introduce and develop the environmental management component in the context of vector and disease control. Attached to the panel are 13 collaborating centres. These are research institutes working in the field of either bio-medical or engineering related topics of tropical disease control. PEEM defines environmental management as all technical and managerial interventions which modify and/or manipulate environmental factors and their interaction with man and vectors.

The overall objectives are to prevent or minimise vector propagation and to reduce man-vector-pathogen contacts in order to achieve an optimal health status for a target population. Chemical vector control would include the application of repellents, attractants, insecticides, molluscicides and chemosterilants. Biological vector control can be achieved by the release of predators and microbial insecticides and by genetic manipulation. Finally, the medical interventions would focus on morbidity control, such as chemotherapy, vaccines, general treatment, improved nutrition, preventive health safeguards, prophylaxis and health education.

Drainage Water Management Measures

In the context of drainage water re-use, treatment and disposal, the potential health risks can be analysed within a simplified system of drainage water flow in combination with the transfer of pathogens, toxic substances and the occurrence of potential vector breeding sites. This allows one to differentiate whether a health risk is related to water quality, vector breeding or both. For each of the three drainage water management options, a number of questions are formulated to identify potential health risks. These are followed by a range of recommendations for possible control and preventive measures.

Direct Re-use

At first glance, re-use may appear to be a water quality problem mainly regarding high salt concentrations which can affect plant growth. However, pesticides in different forms can be washed into surface drains and trace elements can be leached out by subsurface drains. Where drainage water with high loads of pesticides and trace elements is re-used for irrigation, a health risk can occur for irrigation workers or for persons using this water for domestic purposes.

In the case of direct re-use, there are few opportunities for the application of environmental management measures. Once drainage water has entered an irrigation system, it is probable that individuals will come into contact with this water. Therefore, effective monitoring of the quality standards of drainage effluents is most important. It is necessary to develop safeguard strategies in case of unacceptable or dangerous contamination levels.

Agriculture-Forestry System and Solar Evaporators

This system of drainage water management, which aims at a continuous concentration of salt in progressively smaller volumes of water, is less of a health risk, as saline water is unsuitable for drinking.

Control and prevention

— Intermittent drying out of ponds or storage tanks for at least 5-7 days.

— If possible, lining of all facilities to avoid seepage and minimise aquatic growth. In the case of earth lining: regular weed control. Only restricted or no applications of insecticides or molluscicides.

— Restricted accessibility in order to reduce man-water contacts.

— Settlement planning: geographical separation of settlements from ponds and tanks.

— Monitoring of vector breeding activities and water quality.

— Evaluation of options.

Constructed Wetlands

The various physical, chemical and biological treatment processes may require a number of water retention structures such as constructed wetlands. This could lead to new and mostly permanent open water surfaces.

Control and prevention

- Water level fluctuations and intermittent drying out of the wetland area.
- Restricted accessibility.
- Geographical location outside and separated from human settlements.
- Off site: settlement planning and maintenance, housing improvement, personal protection.
- Health education.
- Monitoring: vector populations, water quality, case reporting.
- In the case of high community vulnerability and environmental receptivity: evaluation of options.

Environmental Management Measures

Drainage systems need to be connected either with subsurface or surface drainage canal outlets. In the case of surface collector canals, additional open water surfaces will be introduced. In warm and hot climatic zones, drainage canals include a number of typical features which favour vector breeding, disease transmission and direct pathogen propagation, such as low and irregular flow velocities, low embankment slopes, high seepage, uncontrolled water access, uncontrolled deposition of excreta and aquatic weed growth.

Environmental Modification

Modifications to the drainage system environment would include: the change from open to piped or covered drains; canal lining with concrete in order to increase flow velocities and reduce aquatic weed growth; installation of special structures for cattle crossings and drinking; and boat ramps to protect earth embankments.

Environmental Manipulation

The key elements of environmental manipulation would be flow and water level management measures. Periodical flushing will help to dislodge snails and mosquito larvae if drag forces or shear stresses due to higher velocities exceed certain limits. Water level fluctuations can have distinct control impacts on both snail and mosquito breeding. If the drop is fast enough, snails, larvae and eggs become stranded and die. Intermittent flow, drying out of canals in connection with flushing, and water table fluctuations can also be effective tools to control mosquito or snail breeding.

However, the approach differs for snails and mosquitoes, according to the locally specific population dynamic and vector bionomics. Finally, canal maintenance means weed control and the removal of sediments. Weed control can be done either mechanically, chemically by applying herbicides, or biologically with fish.

Interventions and environmental changes related to human habitation might not be sufficient if canals are continuously misused for excreta and waste disposal. In this case, health risks have to be minimised with a set of non-drainage related measures. This includes the improvement of sanitation facilities and personal hygiene (e.g., water supply systems or latrines) and the planning and maintenance of settlements.

There are no standard packages of engineering techniques available, nor should environmental management be considered as the ultimate solution for controlling water related parasitic diseases. Birley has introduced a systematic approach to forecast vector-borne disease implications. In this assessment methodology he addresses three main components which contribute to the potential health hazards:

Community vulnerability: This describes the prevalence of specific diseases in social groups such as children, adults, males, females, workers or farmers. The prevalence is brought into relation with the proximity of disease foci, the immune status, previous history of exposure, general health status and the role of migrants. Vulnerability is ranked as high, moderate or low.

Environmental receptivity: This is the receptivity to transmission of the pathogen related to the abundance of the vector, to human contacts with water or vectors and to any other ecological or climatic factors favourable for transmission. The assessment is followed according to possible transmission but not to occurrence, transmission easily resumed, or to high receptivity.

Vigilance of the health services: This describes the quality and performance required of a health service to cope with an increased health hazard. The question is whether a health service is able to support and manage vaccination campaigns, continuous case detection, drug provision and delivery, hospitals, sufficient and skilled staff, health education and information or means for chemical or biological vector control. The ranking includes: very good, effective preventive measures only, effective treatment only, and none.

The assessment of control strategy effectiveness is organised in a sequence of flow charts and worksheets. It includes a comprehensive set of questions which finally lead to the ranking of the three components. The methodological approach forces one to focus on control elements and will structure data and help to set up additional survey or monitoring concepts if there is a lack of data. It will also help in assessing the quality and reliability of data. Overall the final outcome (the total assessment of potential health hazards) will provide a sound basis for identifying the required set of safeguards and preventive intervention measures, including the environmental engineering required to control water management related diseases.

9

Drainage Water Disposal

Disposal in to Natural Hydrological Systems

There are a limited number of options available when trying to decide where and how to dispose of agricultural drainage Water into the natural hydrological system. The common option is to return the water either to the land as part of the irrigation water supply, or to rivers and lakes, or to salt sinks, such as the ocean. The options available to any single project may be limited because of water quality concerns. Drainage water quality may vary within a catchment. Frequently, in developing countries, agricultural drains are used for the disposal of domestic and industrial wastewater, or for the disposal of polluted water from other non-agricultural sources. This may adversely affect the quality of agricultural drainage water, and limit its potential for reuse.

Downstream beneficial uses of any surface water body to which drainage water is added must be protected. For example, it may not be appropriate to discharge saline drainage water into a river or lake when that surface water body is being used for domestic or agricultural water supplies. In many cases, it may be possible and fully acceptable to discharge drainage water into a large freshwater body. However, it will be necessary to determine the assimilative capacity of the receiving water and identify the constituents in the drainage water to determine the 'safe level' or discharge requirements for the drainage water.

The discharge requirements should specify the maximum allowable concentration of each constituent of concern and the volume of drainage water discharge that will be acceptable. In some cases, there will be a significant difference between the quality of the drainage water and that of the receiving water, while in other cases, there will be little difference between the two. The dilution capacity of the receiving water will vary from place to place and from time to time depending on numerous local conditions and the upstream uses of the receiving water. The discharge of drainage water of a higher quality than the receiving water is generally acceptable.

Pollutants in the drainage water may end up in the channel bed material. Here, they may create subsequent water quality problems, if and when the channels need to be dredged. Special attention should also be given to the pollution hazards posed by regional seepage flows. This seepage may pollute the receiving surface water with solutes picked up during movement through various soil or rock formations. This may reduce the dilution capacity of surface water. Shallow seepage may cause excessive nutrient loading, while deeper seepage may convey toxic geochemicals. Mitigation to facilitate the disposal of drainage to surface waters should preferably start at the field level.

Measures such as the construction of retention ponds and the establishment of riparian border strips can be taken to reduce the direct inflow of polluted surface drainage water into the receiving system. In the receiving system, chemical concentrations may be further reduced by dilution and mixing. Less polluted or even fresh water from upstream storage can sometimes be used for dilution and flushing, so as to facilitate safe drainage discharge into streams and lakes. The disposal of drainage water from acid sulphate soils is unique. The pH of this drainage water can be as low as three or four. Generally, this type of drainage water must be highly diluted in the main drainage system. Where dilution is not possible, special arrangements should be made to flush and transport this acid drainage water from the system.

In most cases, lakes must be given special consideration, as there may not be an adequate outlet or flow volume from the lake, and contaminants may accumulate. This could cause substantial long-term problems to any beneficial uses of the lake water, such as aquatic habitat.

Generally, rivers continually cleanse themselves and can tolerate somewhat higher pollutant constituent loads than lakes. However, it is still essential that the quality of both the drainage water and the river water be evaluated to determine the safe level of constituents that can be placed in the river without affecting downstream beneficial uses.

Estuaries, bays and oceans have somewhat different water quality requirements. Considerable exchange of water takes place in the estuarine, bay and ocean environments due to the changing tides and ocean currents. In addition, the introduction of saline drainage water into a salt-water environment generally reduces the impact of the discharge of the drainage water on the receiving water. However, each receiving water needs to be analysed and compared with the drainage effluent in order to protect beneficial uses of the receiving water. Nutrients in drainage waters may cause problems in estuaries. If possible, the oceans and seas should be the ultimate natural disposal location for saline drainage water.

In irrigated areas, there is often some low quality land available near to where agricultural drainage problems occur. Such land can be used for the disposal of drainage water in evaporation ponds or constructed wetlands, or for re-use in saline agriculture-forestry. However, if the drainage water contains potentially toxic constituents, such as Se, then disposal of drainage water in evaporation ponds or in wetlands will be environmentally hazardous. In California's Central Valley, drainage water containing Se in concentrations greater than 2 m g/litre is not recommended for use as water for ponds or wetlands. Some proportion of the water must be flushed through such wetlands to prevent toxic ion concentration from accumulating to dangerous levels.

Land retirement is sometimes portrayed as a 'solution' to an agricultural drainage problem where salt is allowed to accumulate on the soil surface due to evaporation from a shallow water table. The principal difficulty with land retirement as a 'solution' to drainage problems is that it takes a considerable amount of land out of production. This is obviously a perplexing issue when farmers lose half of their land (the salt sink area) in order to continue production on the remainder. This concept of taking care of a salt balance-waterlogging problem has recently been termed 'dry drainage'.

The objective is to have the non-cropped area large enough and the evaporation from this area fast enough to achieve the necessary salt balance and a stable water table over the entire area. The theory is that if inflow balances outflow, then the water table will be stable. The cropped area must be continuously irrigated to avoid salt accumulation in the cultivated portion of the fields, since the water table in the irrigated lands will be higher than it is in the adjacent lands. Dry drainage is not recommended.'Dry drainage' adds to the evaporative water loss of the area.

Evaporation Ponds

Disposal of runoff and drainage waters into natural depressions has been practised for centuries. The impounded waters are dissipated by evaporation, seepage and transpiration losses. Use of constructed disposal basins for saline agricultural drainage waters are also common worldwide where there are constraints on discharging into natural salt sinks such as the oceans and inland closed basins.

In the Murray-Darling basin in Australia, some of the constructed evaporation basins are intended to hold saline water only temporarily. The stored waters are then released during high river flows. Closed basins are expected to have operating lives of 50-150 years. The surface areas of these basins range from a few hectares to about 2,000 ha. The waters impounded in the Australian basins are typically dominated by NaCl type salts. Data on trace elements have not been reported. Lateral and vertical seepage losses from these basins are substantial, 20-50% in most basins.

The evaporation basins in the San Joaquin Valley receive about 3.9 million m^3/year of subsurface drainage water from about 22,700 ha of subsurface drained fields. Some ponds or pond cells are allowed to dry, while others are maintained at a minimal water level. The ponds are typically located on clay loam to clay soils. Emergent vegetation is controlled but the submerged rooted vegetation and phytoplankton are not. Rainfall at the evaporation pond facilities is low, from less than 100 to 250 mm/year. The grass ET rate in this region is about 1,500 mm/year.

The hydrology of ponds is comparatively simple. The inputs into the ponds include drainage water from croplands, rainfall and pumped drainage water from perimeter drains installed to intercept pond seepage. The outputs include water evaporation, unrecovered pond seepage and ET from aquatic vegetation. Estimates of pond seepage rates have been made by point measurements as well as by mass balance of pond inflows, changes in water surface elevation and evaporation rates.

Mass-balance based estimates for several different ponds have yielded similar results despite differences in the soil texture of the pond bed material. Although initial seepage rates upon filling a newly constructed pond may be as great as 10 mm/day, seepage rates decrease dramatically in a few months to a few millimetres per day. This decrease in seepage is attributed to the plugging of conducting pores in the bed material by microbial slimes and colloidal soil materials. Seepage losses are generally dependent on local soil, geohydrological and topographical conditions, and excessive seepage could lead to environmental problems.

For example, in Pakistan, seepage from evaporation ponds caused considerable waterlogging and salinisation in surrounding farmland. Therefore, pond site selection is important, and seepage losses should be controlled. A lining and leachate collection system may be required to ensure that there will be no contamination of usable groundwater. The type of lining, clay or synthetic, should be determined by material availability, local requirements or environmental regulations. Evaporation from free water surfaces is influenced by many variables such as air temperature, wind speed, humidity, net solar radiation and water salinity.

The evaporation rates from ponds have been measured with floating evaporation pans and diurnal monitoring of pond waters and nearby terrestrial climatic data. Results from floating evaporation pans containing water of the same salinity in the pond show that evaporation rates from water surfaces decline with increasing water salinities due to a reduction in water surface vapour pressure. For example, pan evaporation rate for an 8-day period in September 1989 was 63.5 mm for water electrical conductivity (EC) of 14 dS/m, 54.5 mm for water EC of 30 dS/m and 52.4 mm for water EC of 47 dS/m.

Evaporation rates have also been estimated using reference evapotranspiration, ETo, data: E=Y(ETo), where Y= 1.3234 - 0.0066 EC (dS/m) for water EC up to 60 dS/m. The measured correction factor, Y, was 1.2 for water EC of about 20 dS/m, 1.07 for water EC of 47 dS/m and 0.92 for water EC of 59 dS/m. Where pond waters are evapoconcentrated to a point where evaporite minerals begin to form, a thin crust of evaporites may form on the water surface during cooler air and water temperatures (night-time), so effectively limiting water evaporation. However, during higher temperatures (daytime), that thin crust of evaporites on the water surface may melt back into solution. With further evapoconcentration, permanent salt deposits will form.

The waters disposed of in evaporation basins or ponds have dissolved mineral salt concentrations ranging from 2,500 to 65,000 mg/litre (m g/g). The estimated annual salt loading into the San Joaquin Valley ponds is about 880 000 t. This is equivalent to about 25 % of the annual salt accumulation in the 0.9 million ha of irrigated lands on the west side of the San Joaquin Valley. The waters impounded in the ponds are $NaSO_4$ or $NaSO_4$/NaCl types. The principal sources of salts are the applied water and the chemical weathering of the marine sedimentary rocks in the Pacific Coast Range and the alluvium formed in the valley floor.

Impounded drainage waters are evapoconcentrated up to about 388,000 mg/litre, more than ten times the 35,000 mg/litre seawater salinity. As the water evapoconcentrates, a sequence of different evaporite minerals form as their solubility products are exceeded. The suite of evaporites deposited is mainly regulated by the initial chemical composition of the disposed water and the evapoconcentration factor. A brine chemistry model has been validated to predict the sequence of evaporite formation with desiccation.

The most common minerals to form in copious quantities during the early stages of evapoconcentration are calcite ($CaCO_3$) and gypsum ($CaSO_4 \cdot 2H_2O$). With further evapoconcentration, glauberite [$Na_2Ca(SO_4)_2$], bloedite [$Na_2Mg(SO_4)_2 \cdot 4H_2O$] and mirabilite ($Na_2SO_4$) are typically formed. The last evaporites to be deposited are halite (NaCl) and thenardite ($Na_2SO_4 \cdot 10H_2O$). Associated with these saline drainage waters are several trace elements of concern, including As, B, Mo, Se, U and V. These trace elements are naturally occurring in the Cretaceous geologic formations of the San Joaquin Valley. Of principal concern is Se.

Selenium was found to be toxic to fish and waterbirds at Kesterson Reservoir in 1980 and 1981, respectively. The influent Se concentration of drainage water from croplands was about 300 m g/litre and had bioaccumulated in the aquatic food chain. Algae and rooted plants bioconcentrated Se by about 560-fold and 600-fold (69 and 73 mg/kg), respectively. Zooplankton and aquatic invertebrates (insects) feeding on algae and rooted plants, respectively, biomagnified Se by about 1.2-fold and 1.7-fold (85 and 122 mg/kg), respectively.

Fish feeding on zooplankton and aquatic insects also biomagnified Se by about 1.5- to 2.2-fold (188 mg/kg). The net result was that the Se bioaccumulation factor from water to fish was 1,540-fold in the Kesterson pond. A similar degree of bioaccumulation of Se took place in waterbirds, but was dependent on whether their food source was benthic or herbaceous. Due to the elevated salinities and harsh environmental conditions prevalent in many pond facilities, comparatively few species of plants, invertebrates and fish are able to tolerate the high water salinity, widely fluctuating water levels, high water temperatures and low dissolved oxygen.

However, widgeon grass (Ruppia maritima), waterboatmen (Trichocorixa reticulata) and brine shrimp (Artemia salina) can become quite abundant. These pond food items are eaten by waterfowl and shorebirds. Selenium becomes toxic to birds when Se substitutes for sulphur in the essential amino acids methionine and cysteine. The hazards of Se to waterfowl may be best evaluated by measuring the Se content in bird eggs and livers. In normal non-contaminated aquatic habitats, the Se concentrations in avian eggs rarely exceed 3 m g Se/g. The likelihood of reproductive impairment of birds increases substantially when sets of eggs contain more than 20 m g Se/g. Arsenic is another constituent of concern.

Despite the presence of elevated concentrations of As in ponds located in the Lake Tulare bed, bird eggs do not accumulate As. Although B has bioaccumulated in bird eggs up to a maximum of 18 m g/g, B alone does not appear to cause toxicity in bird embryos. Evaporation ponds also contain elevated levels of Mo, but Mo concentrations up to 16 m g/g do not have adverse effects on birds. Therefore, Se is the element of greatest concern to waterbirds attracted to evaporation basins in the San Joaquin Valley.

The design and management of evaporation pond facilities to enhance desiccation of impounded waters and reduce hazards to wildlife are important considerations. Numerous drainwater treatment and disposal options were investigated immediately following the discovery of Se toxicosis of waterbirds at Kesterson Reservoir. Among those investigated were: desalination reverse osmosis, microbially-mediated anaerobic reduction of selenate and selenite to elemental Se, microalgal-bacterial removal of Se, biomethylation and volatilisation of Se, adsorption of Se to iron filings, chemical reduction of Se, deep well injection, and drainage water re-use. The removal of trace elements in influent and impounded pond waters is inherently difficult due to the saline matrix and extent of removal required to protect wildlife.

The various biological methods of removing Se in waters are capable of reducing Se to about 20-50 m g/ litre. With additional chemical and physical treatment processes in tandem, reduction to about 10 m g/litre is possible. These levels of treatment are inadequate to protect wildlife from hazardous bioaccumulation of Se. The recommended criterion for Se in waters is 1 m g/litre for aquatic life and this criterion appears to be unattainable using currently known treatment technologies. Some dilution would also be required to achieve acceptable concentrations of Se. The impoundment of drainage waters in evaporation ponds can have an effect similar to lagoon treatment of municipal wastewaters.

Selenium, the principal constituent of concern in evaporation basins, is subject to a number of sink/ dissipation mechanisms within the pond environment. Selenite, but not selenate, is strongly adsorbed by soil materials with exposed hydroxyl groups such as oxides of iron, aluminium and other metals. Inorganic Se may be reduced and immobilised into elemental Se and selenides. Indigenous bacteria, fungi and microalgae can methylate Se to dimethylselenide (DMSe) that may eventually volatilise into the atmosphere. In fact, biomethylation and volatilisation of Se is the principal pathway by which Se enters into the atmosphere in the natural biogeochemical cycle of Se.

Other processes that remove Se from the water column are the uptake of Se by aquatic plants and the subsequent deposition of organic Se into bottom mud. However, the Se removed from the water column into immobilised forms may be regenerated back into the water column by oxidation. Moreover, Se bioaccumulated by aquatic biota may become part of the aquatic food chain and pose a hazard to animals at the uppermost level of the food chain. The only mechanisms by which the Se inventory may be dissipated from ponds are by inadvertent seepage losses or by volatilisation into the atmosphere. The former is not desirable because of the potential contamination of shallow groundwaters. As for the latter, emission of DMSe does not appear to have adverse impacts.

Disposal by Injection

Disposal by injection is a process in which liquids are pumped into a well for placement into porous rock or sand formations below the ground surface. The well is generally called a 'deep well' because the proposed injection occurs beneath the lowermost underground source of usable water. Most formations selected for disposal reservoirs in California are old marine sediments containing concentrated saltwater. Any porous and permeable rock formation such as sandstone can act as a disposal reservoir for the injection liquid.

In California, deep well injection technology has been used for more than 60 years for the subsurface disposal of oil field brines. During 1994, about 74.3 million m^3 of brine waters were disposed of in California oil fields by deep well injection as reported by the California Division of Oil and Gas. Oil field practices all across the United States have established deep well injection as a viable alternative method for the disposal of these types of industrial wastes. Injection well disposal of agricultural drainage water is a viable alternative for disposal where receiving formation conditions are adequate and the costs of injection are not prohibitive.

Prior to formal injection testing, the formation aquifer into which drainage water is to be injected must be sampled and tested to document baseline water quality and to assess chemical compatibility with the drainage water. Drainage water mixed with formation water should be analysed to ensure that the two fluids will not produce precipitates that would clog the injection well. Water from all zones sampled should be analysed for calcium hardness, total alkalinity, TDS, temperature and pH. The results from these analyses will serve as baseline data for later use if monitoring of confining layer containment is necessary.

A suitable formation is usually a sand or sandstone which has sufficient porosity and thickness to receive the injection fluid. Significant space can be created over a large volume of receiving formation, when the formation is compressible. The disposal formation should be extensive enough to offer a sizeable reservoir and should be deep enough to allow adequate injection pressures in the injection wells. The pressure build-up required for injection occurs at the injection well head and at varying distances from the well over a period of time.

The injection fluids begin to displace natural fluids and a resulting pressure rise becomes evident at the well head. This well head pressure rise can be modelled using the Bernard formula for pressure build-up. The Bernard formula relies on certain site specific data such as porosity, permeability, formation thickness and other data which can only be determined from core samples and well head pressure testing. A continuously monitored leak detection system for the injection system must be incorporated into a Class I well design. A Class I well is designed to protect the entire area above the uppermost confining layer from leaks of the injected drainage water.

The monitoring system would automatically shut off the injection pumps and instantly reduce the pressure during a sudden leak. The annular space between the injection tubing and the middle casing is filled with a non-corrosive saline fluid. The saline fluid exerts a specified annular fluid pressure along the column. This annular fluid pressure will remain constant if the tubing and cement plugs are not leaking. Pressure in the annular space can now be monitored during testing of the injection system by an electronic pressure sensor.

The construction of deep injection wells will generally require some form of permit. In the United States, a federal government permit is required from the USEPA pursuant to the Safe Drinking Water Act regulations. The EPA underground injection control programme requirements state that an injection well must be below the lowermost formation that contains (within 400 m) a well used for drinking water and be within a geological formation capable of receiving the liquid.

The EPA has designated underground sources of drinking water at a TDS level of 10 000 mg/litre or less. A Class I well under these requirements covers the use of injecting hazardous and non-hazardous wastewaters that are not considered part of oil field operations. The permit also requires extensive geological, construction, testing procedure, monitoring, well abandonment, and pressure build-up data. The information and findings of all relative data are open to public scrutiny and hearings.

The most important consideration in the siting of an injection well is to locate a suitable geological formation for injection of the drainage water. The formation overlying the injection formation must be highly impermeable so that it will act as a hydraulically confining barrier which will prevent upward migration of the injection fluid. Both formations should be thick enough to ensure the desired injection rate. The confining formation should be located as far away as possible from any known faults.

The next most important consideration in the location of an injection well is distance from the drainage source. Long distance from the injection well site to the source results in added costs for construction, pumping and future maintenance. However, this may be required if no suitable injection formation is available in proximity to the drainage water source.

The receiving formation for injection must have a high permeability, measured in millidarcies (md), to provide for an adequate injection rate. One millidarcy is about 1 mm/day for water. Core samples of the injection zone and confining layer may be analysed in the laboratory by testing the air permeabilities. The liquid permeabilities can be conservatively estimated to be 50% less than the laboratory measured air permeabilities. Porosity or the percentage of voids in a soil or rock sample should be estimated by using an appropriate method such as sonic log data from the open hole logs.

Plugging of Injection Wells

The plugging of injection wells by micro-organisms is a common problem encountered in wells that produce oil. Injecting agricultural drainage water which contains nitrate into a formation that contains organic matter and ferrous iron will produce excellent growth conditions for nitrate reducing bacteria. If the drainage water is untreated before injection, the formation will probably clog as the pores of the sandstone formation become filled by an accumulation of biotite. There are two common chemical treatments used by the oil industry to improve the injection capability of an injection well.

The first method is to continuously add sufficient chlorine to produce a chlorine residual of 0.2-1.0 mg/g. The best time to add chlorine to drainage water is just before injection. This treatment may solve only a small portion of the injection problem. If the sandstone formation has a biological nitrate demand, it may also have a chemical chlorine demand, and will remove chlorine from the chlorinated drainage water. A rapid and complete removal of chlorine from the drainage water may suggest a very reactive reducing agent is present in the formation. The second method for treating an injection well is to add a buffered solution of hydrofluoric acid to the well. Laboratory tests have shown that permeability can be increased, but not in all cases.

10

Saline Water Irrigation

Several physical, chemical and biological soil management measures help facilitate the safe use of saline water in crop production. Some important ones in this regard are: tillage, deep ploughing, sanding, use of chemical amendments and soil conditioners, organic and green manuring and mulching.

Tillage is a mechanical operation that is usually carried out for seedbed preparation, soil permeability improvement, to break up surface crusts and to improve water infiltration. If tillage is improperly executed, it might form a plough layer or bring a salty layer closer to the surface. Sodic soils are especially subject to puddling and crusting; they should be tilled carefully and wet soil conditions avoided. Heavy machinery traffic should also be avoided. More frequent irrigation, especially during the germination and seedling stages, tends to soften surface crusts on sodic soils and encourages better stands. Deep ploughing refers to depths of ploughing from about 40 to 150 cm. It is most beneficial on stratified soils having impermeable layers lying between permeable layers.

In sodic soils, deep ploughing should be carried out after removing and reclaiming the sodicity, otherwise it will cause complete disturbances and collapse of the soil structure. Deep ploughing to 60 cm loosens the aggregates, improves the physical condition of these layers, increases soil-water storage capacity and helps control salt accumulation when using saline water for irrigation. Crop yields can be markedly improved by ploughing to this depth every three or four years.

The selection of the right plough types (shape and spacings between shanks), sequence, ploughing depth and moisture content at the time of ploughing should provide good soil tilth and improve soil structure. Special equipment can even invert whole soil profiles or break up substrata as deep as 2.5m that impede deep percolation, so that many adverse physical soil conditions associated with land irrigated with saline water can be modified in order to improve leachability and drainability.

Sanding is used in some cases to make a fine textured surface soil more permeable by mixing sand into it, thus a relatively permanent change in surface soil texture is obtained. When properly done, sanding results in improved root penetration and better air and water permeability which facilitates leaching by saline sodic water and when surface infiltration limits water penetration. The method can be combined with initial deep ploughing.

Chemical amendments are used to neutralise soil reaction, to react with calcium carbonate and to replace exchangeable sodium by calcium. This decreases the ESP and should be followed by leaching for removal of salts derived from the reaction of the amendments with sodic soils. They also decrease the SAR of irrigation water if added in the irrigation system. Gypsum is by far the most common amendment for sodic soil reclamation, particularly when using saline water with a high SAR value for irrigation. Calcium chloride is highly soluble and would be a satisfactory amendment especially when added to irrigation water.

Lime is not an effective amendment for improving sodic conditions when used alone but when combined with a large amount of organic manure it has a beneficial effect. Sulphur too can be effective; it is inert until it is oxidised to sulphuric acid by soil microorganisms. Other sulphur-containing amendments (sulphuric acid, iron sulphate, aluminium sulphate) are similarly effective because of the sulphuric acid originally present or formed upon microbial oxidation or hydrolysis.

The choice of an amendment for a particular situation will depend upon its relative effectiveness judged from its improvement of soil properties and crop growth, the availability of the amendments, relative cost involved, handling and application difficulties and time allowed and required for the amendment to react in soil and effectively replace adsorbed sodium. Attempts have been made to coagulate soil particles and provide deep aeration and better permeability and water infiltration by chemical treatment. Treating the soil with dilute bituminous emulsions can result in effective aggregation, improved aggregate stability and reduced surface crust formation.

Water percolation rate is faster in bitumen-treated soil. Sulphate lignin conditioner can also be used to improve soil structure, and to improve soil permeability. Soil conditioners can have practical applications in seedling establishment when soil is irrigated with saline water of high SAR. Stability of soil aggregates prevents dispersion and formation of deposit crusts and infiltration can be maintained by application of small quantities of organic polyelectrolytes to the soil surface. They can be effective when introduced in the irrigation water or when sprayed over the soil surface.

Mineral fertilisers: Salt accumulation affects nutrient content and availability for plants in one or more of the following ways: by changing the form in which the nutrients are present in the soil; by enhancing loss of nutrients from the soil through heavy leaching or, as in nitrogen, through denitrification, or by precipitation in soil; through the effects of non-nutrient (complementary) ions on nutrient uptake; and by adverse interactions between the salt present in saline water and fertilisers, decreasing fertiliser use efficiency. Crop response to fertiliser under saline or sodic conditions is complex since it is influenced by many soil, crop and environmental factors.

The benefits expected from using soil management measures to facilitate the safe use of saline water for irrigation will not be realised unless adequate, but not excessive, plant nutrients are applied as fertilisers. The level of salinity may itself be altered by excess fertiliser application as mineral fertilisers are for the most part soluble salts. The type of fertiliser applied, when using saline water for irrigation, should preferably be acid and contain Ca rather than Na taking into consideration the complementary anions present. Timing and placement of mineral fertilisers are important and unless properly applied they may contribute to or cause a salinity problem.

Organic and green manures and mulching: Incorporating organic matter into the soil has two principal beneficial effects of soils irrigated with saline water with high SAR and on saline sodic soils: improvement of soil permeability and release of carbon dioxide and certain organic acids during decomposition. This will help in lowering soil pH, releasing calcium by solubilisation of $CaCO_3$ and other minerals, thereby increasing ECe and replacement of exchangeable Na by Ca and Mg which lowers the ESP. Growing legumes and using green manure will improve soil structure. Green manure has a similar effect to organic manure.

Salinisation during fallowing may be severe where a shallow water table exists, since evaporation rates of about 8, 3 and 1 mm/day could occur from the dry surface of fine sandy loam when the water table is kept at 90, 120 and 180 cm from the soil surface, respectively. Mulching to reduce evaporation losses will also decrease the opportunity for soil salinisation. When using saline water where the concentration of soluble salts in the soil is expected to be high in the surface, mulching can considerably help leach salts, reduce ESP and thus facilitate the production of tolerant crops. Thus, whenever feasible, mulching to reduce the upward flux of soluble salts should be encouraged.

Delivery and Distribution of Water

Water delivery and distribution systems must be operated efficiently to facilitate the timely supply of water in the right quantities and to avoid waterlogging and salinity build-up in irrigated lands, especially when saline waters are involved. The amount of water applied should be sufficient to supply the crop and satisfy the leaching requirement but not enough to overload the drainage system. Over-irrigation contributes to the high water table, increases the drainage requirement and is a major cause of salinity build-up in many irrigation projects of the world. Therefore, a proper relation between irrigation, leaching, and drainage must be maintained in order to prevent irrigated lands from becoming excessively waterlogged and salt-affected.

Often irrigation water delivery and distribution systems are over-designed, in the absence of reliable data or appropriate methods to predict project water requirements. It is all the more important, when using saline waters, that excessive amounts are not diverted into irrigation schemes as this is likely to cause more damage than excessive amounts of "good quality" water. FAO has developed methods to determine project water requirements based on actual crop water needs, leaching requirements and irrigation efficiencies. A computer program, called CROPWAT has been developed to calculate crop water requirements and irrigation requirements from climatic and crop data.

Further, the program allows the development of irrigation schedules for different management conditions and the calculation of scheme water supply for varying cropping patterns. The program runs on any standard personal computer with a minimum of 360 Kb of memory. The program can be obtained from FAO on request. A complementary computerised database program called CLIMWAT is available to obtain the required climatic data for CROPWAT. CLIMWAT has data from a total of 3262 meteorological stations from 144 countries. Excessive loss of irrigation water from canals constructed in permeable soil is a major cause of high water tables and secondary salination in many irrigation projects.

Such seepage losses should be reduced by lining the canals with impermeable materials or by compacting the soil to achieve a very low permeability. Because the amount of water passing critical points in the irrigation delivery system must be known in order to provide water control and to achieve high water-use efficiency, provisions for effective flow measurement should be made. Unfortunately, many current irrigation systems do not use flow measuring devices and, thus, the farmers operate with limited control and knowledge of the amount of water actually diverted to the farms.

In addition, many delivery systems encourage over-irrigation because water is supplied for fixed periods, or in fixed amounts, irrespective of seasonal variations in on-farm needs. Salinity and water table problems are often the result. The distribution system should be designed and operated so as to provide water on demand and in metered amounts as needed to achieve high efficiency and to facilitate salinity control and the use of saline waters for irrigation.

Improvements in salinity control generally come hand-in-hand with improvements in irrigation efficiency. The key to the effective use of saline irrigation waters and salinity control is to provide the proper amount of water to the plant at the proper time. The ideal irrigation scheme should provide water as nearly continuously as possible, though not in excess, as needed to keep the soil water content in the rootzone within optimum safe limits. However, carefully programmed periods of stress may be needed to obtain maximum economic yield with some crops; cultural practices also may demand occasional periods of dry soil.

Thus, the timing and amount of water applied to the rootzone should be carefully controlled to obtain good water use efficiency and good crop yield, especially when irrigating with saline water. As mentioned above, this requires water delivery to the field on demand which, in turn, requires the establishment of close coordination between the farmer and the entity that distributes the water; it calls for the use of feedback devices to measure the water and salt contents and potentials in the soil and devices to measure water flow (rates and volumes) in the conveyance systems. The method and frequency of irrigation and the amount of irrigation water applied may be managed to control salinity.

The main ways to apply water are basin flooding, furrow irrigation, sprinkling, subirrigation, and drip irrigation. Flood irrigation is good for salinity control when using saline waters if the land is level, though aeration and crusting problems may occur. Aeration and crusting problems are minimised by using furrow irrigation, but salts tend to accumulate in the beds. If excess salt does accumulate, a rotation of crops and periodic irrigation by sprinkler or flooding should be used as salinity-control measures. Alternatively, cultivation and irrigation depths should be modified, once the seedlings are well established, to "shallow" the furrows so that the beds will be leached by later irrigations.

Irrigation by sprinkling may give better control of the amount and distribution of water; however, the tendency is to apply too little water by this method, and leaching of salts beyond the rootzone may sometimes be accomplished only with special effort. Salinity can be kept low in the seedbed during germination with sprinkler-irrigation, but crusting may be a problem. Emergence problems associated with such crusting may be overcome with frequent light irrigations during this time or by use of special tillage techniques. Subirrigation with saline water is not generally advisable unless the soil is periodically leached of the accumulated salts by rainfall or by surface applications of low-salinity water.

Drip irrigation, if properly designed, is recommended for use of saline irrigation water because it minimises salinity and matric stresses in the rootzone, though salts accumulate in the periphery of the wetted area. Higher levels of salinity in the irrigation water can be tolerated with drip as compared with other methods of irrigation. Because soluble salts reduce the availability of water in almost direct proportion to their total concentration in the soil solution, irrigation frequency should be increased so that the moisture content and salinity of irrigated soils are maintained as high and low, respectively, as is practicable, especially during seedling establishment and the early stage of vegetative growth, if it can be done without resulting in excessive leaching or insufficient depth of rooting.

The most practical way to accomplish this is through use of drip irrigation. Additional water (over that required to replenish losses by plant transpiration and evaporation) must be applied, at least occasionally, to leach out the salt that has accumulated during previous irrigations. This leaching requirement depends on the salt content of the irrigation water and on the maximum salt concentration permissible in the soil solution which depends in turn on the salt tolerance of the crop and the manner of irrigation.

Fortunately, much of the needed leaching can be achieved between crops or during pre-irrigation and early growth-stage irrigations when soil permeability is generally relatively high, especially when using low-salinity waters in the cyclic use strategy. The first irrigations provided for the renewal of cropping following a fallow or uncropped period often unavoidably result in relatively high leaching. Many irrigation practices, especially with flood irrigation, inadvertently result in excess leaching, especially during pre-plant or early-season irrigations before the soil aggregation has slaked and surface soil permeability has diminished.

Effects of non-uniform crop stand and cover, soil infiltration rates (permeabilities) and water application and distribution result in generally non-uniform leaching across an irrigated field. Calculation of the leaching requirement is disproportionately subject to errors related to uncertainties in knowledge of evapotranspiration, since $L_{r,} = 1 - V_{et}/V_{iw}$. The value, much less the distribution, of evapotranspiration is not precisely known for most field situations, especially for conditions of irrigation with saline waters and in the presence of shallow, saline water tables.

Consequently, there is little documented evidence of the positive benefits of increased leaching on crop yield under actual field conditions when irrigating with saline waters. While, certainly, the excess salts applied with saline irrigation waters must be removed over time to sustain crop production, for both short- and long-season crops it is generally sufficient to intentionally apply extra water for leaching only if and when the levels of salinity in the active rootzone actually become excessive. Giving extra water for leaching according to traditional $L_{r,}$. equations with each and every irrigation is not necessary.

Rainfall in sub-humid climates often provides the required leaching. The control of salinity by leaching is accomplished most easily in permeable coarse-textured soils. Medium- and fine-textured soils have the agronomic advantage of a greater water-holding capacity and ordinarily present no major problem from the standpoint of irrigating with saline water and salinity control, particularly if they have good structure and are underlain by a sand or gravel aquifer which facilitates the removal of drainage water. Prevention of excessive salt accumulation is generally more difficult in fine-textured, stratified and slowly permeable soils.

Automated solid-set and centre-pivot sprinklers systems are conducive to good control and uniform distribution of applied water; in principle, trickle irrigation 5s even better. But gravity systems, if designed and operated properly, can also achieve good uniformity. Precision land grading and use of smaller water applications should be used to facilitate the achievement of high uniformity of areal water distribution over the field and infiltration, respectively. Closed conduits, rather than open waterways should be used for water distribution laterals if possible; they have the advantage of more effective off-on control, in addition to capturing gravitational energy for use in pressurising delivery systems or controls which offer better potential for achieving high irrigation efficiencies.

The most advanced centre pivot irrigation system now used by some farmers in the USA is called the LEPA system—Low Energy Precision Application. In this system the sprinkler and spray nozzles used on the centre pivot systems are replaced with drop-down hoses and low pressure emitters that operate at only 0.3 kg/cm^2 and are placed as close as 20-45 cm above the ground. Experience has shown that these systems can reduce evaporation and wind-drift losses to less than 5 percent of the emitted water. The new LEPA technology became commercially available in 1986 and since then has been adopted by an estimated 1000 farmers.

The advantage of LEPA with regard to saline water irrigation is that it can irrigate crops with the right amount of water, avoiding excess and runoff, and minimise foliar damage which is common with saline water irrigation. However, the technique is new, costly and needs to be further developed to reduce costs and make the system simpler for adoption by a wider group of farmers. In furrow-irrigated areas, furrow length should be reduced in order to improve intake distribution and to reduce tail water runoff. Worstell's multi-set system is useful for such purposes. Surge irrigation techniques can sometimes be used to improve irrigation uniformity in graded furrows.

For tree crops, a low-head bubbler system can be used to provide excellent control and to minimise the pressure requirements and expensive filtration systems. Drip systems, of course, are increasingly being used for permanent crops and high-value annual crops and are well suited for use with saline irrigation waters. All opportunities to modify existing irrigation systems to increase their effectiveness of water and salinity control should be sought and implemented. Irrigation management technology for salinity control is the subject of reviews by van Schilfgaarde; van Schilfgaarde and Rawlins and Kruse et al..

A frequent constraint in improving on-farm water use is the lack of knowledge of just when an irrigation is needed and of how much capacity for storage is available in the rootzone. Ways to detect the onset of plant stress and to determine the amount of depleted soil water are prerequisites to supplying water on demand and in the amount needed. Prevalent methods of scheduling irrigation usually do not, but should, incorporate salinity effects on soil-water availability. When irrigating with saline waters, the osmotic component of the soil water potential of the rootzone must be considered in scheduling decisions.

Ideally, irrigation management should have the available soil water near the upper limit during germination and emergence but depleted by about 50 percent, or more, at harvest and should maintain available water within the major rootzone during the early vegetative, flowering and yield formation growth stages at a level which produces no deleterious plant water stress through successive, properly-timed irrigations.

Under saline conditions, some "extra" water must be given for leaching—a minimum commensurate with salt tolerance of the crop being grown, if rainfall is inadequate in this regard. Some method of assessing the water availability to the crop with sufficient lead time to provide for a water application before significant stress occurs should be used for irrigation scheduling purposes. In addition, the amounts of water needed for replenishment of the depleted soil moisture from the rootzone and for leaching must be determined. Prevalent methods used to determine the onset of stress include both direct and indirect measurements.

Leaf water potential can be measured with a pressure "bomb" and used to determine the onset of stress; however, the method does not give information with which to predict when the stress will occur much in advance of its occurrence, nor does it provide a measure of the amount of water to apply. Infrared thermometry can also be used to measure plant water stress indirectly which results in the partial closure of leaf stomates and in reduced transpiration rates, causing leaf temperature to rise abnormally above ambient air temperature. This temperature difference can be interpreted in terms of a crop water stress index with which irrigation-need can be assessed. It suffers the same limitation as the leaf water potential method.

Various scheduling methods can be used which are based on sensing depletion of soil-water per se or soil water potential (matric, osmotic or total), or some associated soil or water property, and knowledge of the critical level (the set-point value). Such levels can be ascertained from salt tolerance data by converting threshold ECe values to osmotic potentials and assuming equivalent crop yield loss (also ET loss) would result from total water potential (i.e. assuming the effects of matric and osmotic potentials are equivalent and additive). Matric potential should be measured by any suitable means. Osmotic potential should be determined by one of the methods of salinity measurement described in Rhoades.

Daily potential evapotranspiration can be calculated from measurements of air temperature, humidity, solar radiation and wind or of pan evaporation. The actual evapotranspiration (ETa) can then be estimated from empirically determined, crop coefficients as described by Doorenbos and Kassam. The summation of these daily ETa values can then be used to estimate accumulative soil water depletion and total water potential. A plot of depletion or water potential versus time is then used to project the need for irrigation. This basic approach can be used based regardless of whether direct measurements of soil water content, or a related parameter, using neutron meters, resistance blocks, time-domain reflectometric (TDR) sensors, four-electrode sensors, or various soil matric potential sensors, etc., are used or estimated from ET methods.

All of the methods suffer the limitation of needing to know the critical set-point value for irrigation, which varies with crop type, rooting characteristics, stage of plant growth, soil properties and climatic stress, etc. An estimate of this value can be obtained as described above or by the method of Doorenbos and Kassam. For saline water, irrigations should be scheduled before the total soil water potential (matric plus osmotic) drops below the level (as estimated above) which permits the crop to extract sufficient water to sustain its physiologic processes without loss in yield.

Since, typically, the crop's root system normally extracts progressively less water with increasing soil depth, the frequency of irrigations should be determined by the level of total soil water potential in the upper half of the rootzone where the rate of water depletion is greatest. Besides the extent of soil water depletion by ET, determination of the amount of water to apply should also be based on stage of plant development, the salt tolerance of the crop at this stage and the status of the soil water salinity at deeper depths in the rootzone.

In early stages of plant development it is often desirable to irrigate just sufficiently to bring the soil to "field capacity" to the depth of present-rooting or just beyond. Eventually, however, excess water must be applied to leach salts accumulated in the upper profile to deeper depth in order to provide the growing plant access to more "usable" soil water in accordance with its expanding needs. Thus, the amount of irrigation water required is dictated by the plant's need for water, the volume of soil reservoir in need of replenishment and the level of soil salinity in the lower rootzone. Benefits of different amounts of saline irrigation water should be determined by evaluating their effects on relative crop yield using the water production function model.

Monitoring Soil Water and Salinity

"Feedback" information on the status of salt and water within the crop rootzone and the extent of leaching being achieved should be obtained periodically to identify developing problem areas, to evaluate the appropriateness of model predictions and as a guide to monitor the effectiveness of the irrigation system and management strategies being used. Soil water content (or matric potential), salinity (and hence osmotic potential) and leaching fraction can, in theory, all be determined from measurements of soil electrical conductivity, ECa, since ECa is a measure of both soil water content and soil water salinity.

Soil salinity in irrigated agriculture is normally low at shallow soil depths and increases through the rootzone. Thus measurements of EC, in shallow depths of the soil profile made over an irrigation cycle are relatively more indicative of changing soil water content (permitting estimation of matric potential), while measurements of ECa in deeper depths of the profile, where less water uptake occurs, are relatively more indicative of salinity. Thus, in principle, depletion of soil water to a set-point level, depth of water penetration from an irrigation or rainfall event and leaching fraction can all be determined from ECa measurements made within the rootzone over time.

However, separate measurements of soil water content and soil water salinity, from which the total water potential can be estimated (matric plus osmotic), are more ideally suited for these needs. The use of time domain reflectometric (TDR) sensors offer potential in this regard. Proper operation of a viable, permanent irrigated agriculture, which also uses water efficiently, requires periodic information on soil salinity, especially with use of saline waters. Only with this information can the effectiveness of irrigation project operation be assessed with respect to the adequacy of leaching and drainage, salt balance and water use efficiency.

Monitoring programs should be implemented to evaluate the appropriateness of model predictions, the effectiveness of control programs, and to assess the adequacy of the irrigation and drainage systems on a project-wide basis. Frequently used methods based on "salt-balance" calculations are inadequate in this regard, for reasons given elsewhere. The direct inventorying and monitoring of soil salinity which are appropriate and needed in this regard are complicated by salinity's spatial variability, since numerous samples are needed to characterise an area.

Monitoring is also complicated by salinity's dynamic nature, due to the influence of changing weather patterns, management practices, water table depth, etc. When the need for repeated measurements is multiplied by the extensive requirements of a single sampling period, the expenditures of time and effort with conventional soil sampling procedures increase proportionately. Hence, simple, practical methods for measuring or predicting field salinity are needed. Procedures for delineating representative areas within irrigation projects, where periodic measurements can be made for monitoring, are also needed, as are procedures for rapidly producing soil salinity maps.

For these reasons new instruments for measuring soil electrical conductivity should be used and coupled with mobile transport vehicles, remotely sensed imagery and computer mapping techniques into an integrated system for inventorying and monitoring soil salinity. These procedures should also be integrated with solute-transport models to develop a geographic information system for salinity assessment and management needs. A network of representative soil salinity monitoring stations should be established in irrigation projects, especially those projects where saline waters are used for irrigation.

Methods for Control Salinity

Integrated strategy to facilitate the use of saline waters for irrigation, to minimise drainage disposal problems and to maximise the beneficial use of multiple water sources

Irrigated agriculture is a major contributor to the salinity of many surface- and groundwaters. The agricultural community has a responsibility to protect the quality of these waters. It must also maintain a viable, permanent irrigated agriculture. Irrigated agriculture cannot be sustained without adequate leaching and drainage to prevent excessive salinisation of the soil, yet these processes are the very ones that contribute to the salt loading of surface and groundwaters. But surface and groundwater salinity could be reduced if salt loading contributions from the irrigation processes were minimised or eliminated.

The protection of water resources against excessive salination, while sustaining agricultural production through irrigation, requires the implementation of comprehensive land and water use policies that incorporate the natural processes involved in the soil-plant-water and associated geohydrological systems. Strategies to consider in coping with increasing salinity in receiving water systems resulting from irrigation include:

— eliminating irrigation of certain polluting lands;

— intercepting point sources of drainage return flow and diverting them to other uses;

— reducing the amount of water lost in seepage and deep percolation;

— isolating saline drainage water from good quality water supplies.

Minimising Deep Percolation and Intercepting Drainage

Minimising leaching always reduces the salt discharged from the rootzone. Additionally, deeply percolating water often displaces saline groundwater of higher salinity or dissolves additional salt from the subsoil. Reducing deep percolation will generally reduce the salt load returned to the river as well as reduce water loss. The "minimised leaching" concept of irrigation which reduces deep percolation should be adopted and implemented to reduce salinisation of water resources associated with irrigation projects, especially in projects underlain by salt-laden sediments. In addition, saline drainage water should be intercepted.

Intercepted saline drainage water can be desalted and reused, disposed of by pond evaporation or by injection into some isolated deep aquifer, or it can be used as a water supply where use of saline water is appropriate. Desalination of agricultural drainage waters for improving water quality is not generally economically feasible even though it is to be implemented for the return flow of the Wellton-Mohawk irrigation project of Arizona, USA. The high costs of the pretreatment, maintenance, and power are deterrents. Only in extreme cases, or for political rather than technical reasons, is desalination advocated.

Isolating and Reusing Drainage Water

While there is an excellent opportunity to reduce the salt load contributed by drainage water through better irrigation management, especially through reductions in seepage and deep percolation, there are practical constraints which limit such reductions. But the ultimate goal should be to maximise the utilisation of the irrigation water supply in a single application with minimum drainage. To the extent that the drainage water still has value for transpirational use by a crop of higher salt tolerance, it should be used again for irrigation.

Drainage waters are often returned by diffuse flow or intentional direct discharge to the watercourse and automatically "reused." Dilution of return flows is often advocated for controlling water salinity. This concept has serious limitations when one considers its overall effect on the volume of usable water, and it should not be advocated as a general method of salinity control. The preferred strategy to control the salinity of water resources associated with irrigated lands is to intercept drainage waters before they are returned to the river and to use them for irrigation by substituting them for the low-salinity water normally used for irrigation at certain periods during the irrigation season of certain crops in the rotation.

When the drainage water quality is such that its potential for reuse is exhausted then this drainage should be discharged to some appropriate outlet. This strategy will conserve water, sustain crop production and minimise the salt loading of rivers that occurs by irrigation return flow. It will also reduce the amount of water that needs to be diverted for irrigation. Integrated strategy to facilitate the use of saline waters for irrigation, to minimise drainage disposal problems and to maximise the beneficial use of multiple water sources

The ultimate goal of irrigation management should be to minimise the amount of water extracted from a good-quality water supply and to maximise the utilisation of the extracted portion during irrigation use, so that as much of it as possible is consumed in transpiration (hence producing biomass) and as little as possible is wasted and discharged as drainage. Towards this goal, to the extent that the drainage water from a field or project still has value for transpirational use by a crop of higher salt tolerance, it should be used again for irrigation before ultimate disposal.

Unavoidable excessive, usable resulting drainage water should be intercepted and isolated from good-quality water supplies and used within dedicated parts of the project as a substitute for part of the freshwater given to the crops. The "dual rotation, cyclic" management strategy of Rhoades can be used to enhance the feasibility of reusing such saline drainage waters for irrigation. In this system, sensitive crops (such as lettuce, alfalfa, etc.) in the rotation are irrigated with "low salinity" water (usually the developed water supply of the irrigation project), and salt-tolerant crops (such as cotton, sugarbeets, wheat, etc.) are irrigated with saline drainage water or the shallow groundwater created by over-irrigation in the project.

For the salt-tolerant crops, the switch to saline water is usually made after seedling establishment; preplant irrigations and initial irrigations being made with low-salinity irrigation water. The secondary drainage resulting from such re-use should also be isolated and used successively for crops (including halophytes and tolerant trees) of increasingly greater salt tolerance. The ultimate unusable drainage water should be disposed of to some appropriate outlet or treatment facility. The feasibility of this "dual-rotation, cyclic" strategy is supported by the following:

— The maximum possible soil salinity in the rootzone resulting from continuous use of saline water does not occur when this water is used only for a fraction of the time.

- Alleviation of salt build-up resulting from irrigation of salt-tolerant crops with the saline water occurs later when a salt-sensitive crop (s) is irrigated with the low-salinity water supply, or during off season periods of high rainfall.
- Proper preplant irrigation and careful irrigation management undertaken during germination and seedling establishment are made using the low-salinity water supply to leach salts accumulated from saline irrigations out of the seed-area and from shallow soil depths.
- Data obtained in modelling studies and in field experiments support the credibility and feasibility of this "cyclic" reuse strategy.

There are many different situations where the use of saline water for irrigation in the recommended strategy could be practical. One situation is where high quality water is available during the early growing season but is either too costly or too limited in supply to meet the entire seasons requirements. This situation is common in parts of India and Pakistan, for example. Where high-quality water costs are prohibitive, crops of moderate to high salt tolerance could be irrigated with saline drainage or groundwater, especially at later growth stages with economical advantage, even if this practice resulted in some reduction in yield relative to that obtainable with a full supply of fresh water.

Use of saline water for irrigation reduces the amount of high-quality water required to grow crops and hence expands the water-resource base for crop production. Another situation conducive for such reuse is one where drainage water disposal, or a means of lowering an excessively shallow water table, is impractical due to physical, environmental, social or political factors. Reuse of the drainage water for irrigation in this situation decreases the volume of drainage water requiring disposal or treatment, and the associated costs. Furthermore, a reduction in the drainage volume also reduces the salt loading of the receiving water.

A difficulty in adopting the cyclic, "dual-rotation" strategy may exist on small farms where the drainage water produced on-site is too little or does not coincide with peak crop-water demand. One method of collecting sufficient quantities of drainage water is to install a network of interceptor drains in areas with shallow water tables. A submersible pump could be placed in collector sumps as a means to access the drainage water. The size of the area that can be irrigated using such "drainage water" will vary, of course, depending on the capacity of the drainage system.

To surface irrigate effectively, at least 10 litres/min/ha is required. Another way to collect drainage water is to install a network of shallow wells in strategic shallow water table areas. The wells can be connected to a common drainage manifold to facilitate collection and distribution. Consultation with irrigation and drainage engineers is advised before installing any drainage water collection system for irrigation use.

Regional management permits reuse in dedicated areas so as to avoid the successive increase in concentration of the drainage water that would occur if the reuse process were to operate on the same water supply and same land area (i.e. in a closed loop). With regional management, certain areas in the region can be dedicated to reuse while other areas such as upslope areas, are irrigated solely with high quality water as usual. The second-generation drainage water from the primary reuse area is discharged to other dedicated reuse areas where even more salt-tolerant crops are grown, or to regional evaporation ponds or to treatment plants.

Ideally, regional coordination and cost-sharing among growers should be undertaken in such a regional reuse system. A novel means of "treating" saline waste waters before their ultimate disposal is to use them to irrigate specific crops that have the ability to accumulate large quantities of undesirable constituents (e.g. Se, Mo, NO_3, B, etc.) in the plants, in order to help reduce adverse ecological effects of disposal. The feasibility of biofiltration, the term used to describe this process, has been demonstrated by Cervinka et al., and Wu et al.. They found that mustard, some grasses and certain native plant species found in California are effective in accumulating substantial amounts of Se in their shoots. This alternative "reuse" practice is most attractive where:

— drainage disposal problems exist related to a potentially toxic trace constituent,

— a bioaccumulator with economic value exists, and

— other treatment processes are either unavailable or too expensive.

An alternative reuse strategy that is often advocated is to blend water supplies before or during irrigation. Blending may be appropriate provided the drainage or shallow groundwater is not too saline per se for the crop to be grown. If the blending strategy is adopted, there must be a controlled means of mixing the water supplies. Shalhevet and Meiri et al. described two blending processes (i) network dilution, and (ii) soil dilution. With network dilution, water supplies are blended in the irrigation conveyance system. With soil dilution, the soil acts as the medium for mixing water of different qualities. A network blending system must be designed and installed if the blending strategy is to be adopted. The theory and design of dilution control systems and their use in irrigation networks has been developed by Sinai et al..

The cyclic strategy is preferred over the blending strategy in that

(i) more salt sensitive crops can be included in the rotation,

(ii) a blending facility is not required, and

(iii) there is less danger in losing "usable water" for the crop.

However, the cyclic strategy will require larger quantities of drainage water during the irrigations where it is used (since the water is not blended) and thus a storage system may be required in order to supply sufficient water for an effective irrigation.

Soil salinity under the cyclic strategy will fluctuate more, both spatially and temporally than in soils irrigated with conventional water supplies. Therefore, predicting plant response will be more difficult under these conditions. Hence, long-term effects on soil salination should be monitored using the techniques described earlier. Management must be adjusted to keep the average rootzone salinity levels within acceptable limits for the crop being grown, considering its stage of growth. Many saline waters contain certain elements, such as boron and chloride, that can potentially accumulate in plants, especially woody, perennial ones, to levels that cause foliar injury and a subsequent reduction in their yield.

In such cases, toxicity may produce more long-term detrimental effects than does salinity. Since boron is adsorbed by the soil it requires longer to build to toxic levels in the soil solution and it requires more leaching to remove its excessive accumulations than does salinity. Thus long-term accumulation in the soil of potential toxicants must be considered, since toxic effects may not become evident for years and may be more difficult to eliminate. Water containing excessive concentrations of B or Cl should not be used to irrigate perennial crops, since,

- these crops are generally more sensitive to specific-ion effects
- they represent a long-term investment, and
- they will have a long time opportunity to accumulate toxic levels.

This same concern applies to growing perennial crops in the presence of a shallow groundwater that contains solutes potentially toxic to the plant.

Another consideration as regards use of saline water for irrigation is the potential of the plant to accumulate certain elements (such as Se, Mo, heavy metals) that are toxic to consumers of the crops (humans and animals).

Since plants vary in their ability to absorb and translocate toxic elements, crops that accumulate large quantities of toxic elements in the edible animal organs should also be avoided when using saline waters containing such elements for irrigation. Fleming found that Se concentrations were higher in Cruciferae (cabbage, cress, radish, rape and turnips), Liliacea (onion), and Leguminosae (clover and peas) than in Compositae (artichoke and lettuce), Gramineae (barley, oats, rye grass, and wheat), and Umbelilferae (parsnip and carrots) when grown on seleniferous soils in Ireland. It is also important to understand how the toxic constituent is partitioned within the plant. In most annual fruit and vegetable crops selenium accumulates more in the leaves than in the fruit, but exceptions exist. If the saline water in question contains high levels of a potentially toxic element, the user should obtain expert advice.

Principles and practices for Safe Use of Saline Water

While irrigated agriculture has greatly increased crop productivity, inappropriate and inefficient irrigation has wasted water, polluted surface water and groundwater, damaged productivity and altered the ecology of vast areas of land. Contamination of water supplies by irrigation is, in many places, posing health risks and drastically increasing the costs of treating waters for domestic and industrial uses. Surface and groundwaters in many areas are being contaminated by salts, fertilisers, herbicides and pesticides.

Toxic chemicals are rendering many developed water supplies unfit for drinking and even for irrigation in some cases. These pollutants also degrade the recreational use and esthetic value of surface waters. At the same time, costly limitations are being placed upon irrigation to reduce its pollutional discharges or to treat its wastes before discharge. Finding a suitable, acceptable place for such discharge is increasingly becoming a major problem in some situations, especially in the developed countries. Blending saline and fresh waters reduces the potential usability of the total water supply.

Use of polluted waters for irrigation limits crop production potential, as well as posing some potential health hazards to the consumers of the food. To overcome the above-described problems, new techniques need to be developed and implemented to reduce excessive water uses and to conserve limited water supplies and better ways must be found to implement existing methods more effectively. Efficiency of irrigation must be increased by the adoption of appropriate management strategies, systems and practices and through education and training. Reuse of wastewater, including the use of drainage water and shallow saline groundwater for crop production, must be made an integral component of irrigation water management, water conservation and environmental protection programmes.

Effective salinity control measures must be implemented to sustain irrigated agriculture and to prevent pollution of associated water resources. Such measures must be chosen with recognition of the natural processes operative in irrigated, geohydrologic systems, not just those on-farm, and with an understanding of how they affect the quality of soil and water resources, not just crop production. Some practices can be used to control salinity within the crop rootzone, while other practices can be used to control salinity within larger units of management, such as irrigation projects, river basins, etc.

Additional practices can be used to protect off-site environments and ecological systems—including the associated surface and groundwater resources. The "on-farm" practices usually consist of agronomic and engineering techniques applied by the farmer on a field-by-field basis. The "district-wide" or "larger organisational basis" practices generally consist primarily of engineering structures for water control (both delivery and discharge) and systems for the collection, reuse, treatment and/or disposal of drainage waters. There is usually no single way to achieve safe use of saline water in irrigation.

Many different approaches and practices can be combined into satisfactory saline water irrigation systems; the appropriate combination depends upon economic, climatic, social, as well as edaphic and hydrogeologic situations. Thus, no procedures are given here for selecting "the" appropriate set of practices for different situations. Rather, some important goals, principles and strategies of water, soil and crop management practices that should be considered in the use of saline water for irrigation are presented as guidelines.

Salinity Management

Salinity management constitutes an important aspect of safe use of saline water irrigation. This requires an understanding of how salts affect plants and soils, of how hydrogeologic processes affect salt accumulation, and also of how cropping and irrigation activities affect soil and water salinity. To prevent the excessive accumulation of salt in the rootzone from irrigation, extra water (or rainfall) must, over the long term, be applied in excess of that needed for ET and must pass through the rootzone in a minimum net amount. This amount, in fractional terms, is referred to as the "leaching requirement".

In fields irrigated to steady-state conditions with conventional irrigation management, the salt concentration of the soil water is essentially uniform near the soil surface regardless of the leaching fraction (LF, the fraction of infiltrated water that actually passes through the root-zone) but increases with depth as LF decreases. Likewise, average rootzone salinity increases as LF decreases; crop yield is decreased when tolerable levels of salinity are exceeded. Methods to calculate the leaching requirement and to predict crop yield losses due to salinity effects were described previously.

Once the soil solution has reached the maximum salinity level compatible with the cropping system, at least as much salt as is brought in with additional irrigations must be removed from the rootzone; a process called "maintaining salt balance." To prevent waterlogging and secondary salination, drainage must remove the precipitation and irrigation water infiltrated into the soil that is in excess of crop demand and any other excessive water (surface or subsurface) that flows into the area; it must provide an outlet for the removal of salts that accumulate in the rootzone in order to avoid excessive soil salinisation, and it must keep the water table sufficiently deep to permit adequate root development, to prevent the net flow of salt-laden groundwater up into the rootzone by capillary forces and to permit the movement and operation of farm implements in the fields.

Artificial drainage systems may be used in the absence of adequate natural drainage. They are essentially engineering structures that control the water table at a safe level according to the principles of soil physics and hydraulics. The water table depth required to prevent a net upward flow of water and salt into the rootzone is dependent on irrigation management and is not single-valued as is commonly assumed. Methods to calculate drainage requirements are given elsewhere. The time-averaged level of rootzone salinity is affected by the degree to which the soil water is depleted between irrigations, as well as by the leaching fraction. As the time between irrigations is increased, soil water content decreases as the soil dries, and the matric and osmotic potentials of the soil water decrease as salts concentrate in the reduced volume of water.

Water uptake and crop yield are closely related to the time and depth averaged total soil water potential, i.e. matric plus osmotic. As water is removed from a soil with non-uniform salinity distribution, the total water potential of the water being absorbed by the plant tends to approach uniformity in all depths of the rootzone. Following irrigation, plant roots preferentially absorb water from rootzone depths with high water potential. Normally this means that most of the water uptake is initially from the upper, less saline soil depths until sufficient water is removed to increase the total water stress to a level equal to that in the lower depths. After that water is removed from the deeper, more saline soil depths and the effect of salinity, per se, on crop growth is magnified. This implies that:

- — forms of irrigation that minimise matric stress, such as drip irrigation, can be used to minimise the harmful effects of irrigating with saline water;
- — high leaching fractions can be used to minimise the buildup (hence harmful effects) of high levels of salinity in deeper regions of the rootzone.

The distribution within and the degree to which a soil profile becomes salinised are also functions of the manner of water application, as well as the leaching fraction. More salt is generally removed per unit of leachate with sprinkler irrigation than with flood irrigation. Thus, the salinity of water applied by sprinkler irrigation can be somewhat higher, all else being equal, than that applied by flood or furrow irrigation with a comparable degree of cropping success, provided foliar burn is avoided. The high salt-removal efficiency of sprinkler irrigation may be explained as follows.

Solute transport is governed by the combined processes of convection (movement of solutes with the bulk solution) and diffusion (independent movement of solutes as driven by a concentration gradient); convection is usually the predominant process in flood-irrigated soils. Differential velocities of water flow can occur within the soil matrix because the pore size distribution is typically non-uniform. This phenomenon is called dispersion. It can be appreciable when flow velocity is high and pore size distribution is large; diffusion often limits salt removal under such conditions. Soils with large cracks and well-developed structure are especially variable in their water and solute transport properties because the large "pores" are preferred pathways for water flow, as are earthworm channels, old root holes, interpedal voids, etc.; most of the flow in flooded soils occurs via these "pores". Much of the water and salt in the small and intra-aggregate pores is "bypassed" in flood irrigated soils.

Flow velocity and water content are typically lower in soils irrigated with sprinklers; hence, bypass is reduced and efficiency of salt leaching is increased. Other soil-related processes also affect salt concentration and transport during the irrigation and leaching of soils. In most arid land soils, the clay particles are dominated by negative charges, which can retard cation transport through adsorption and/or exchange processes. Simultaneously, anions are largely excluded from that part of the pore solution adjacent to the negatively-charged clay surface; this accelerates their relative transport. The borate anion also undergoes adsorption reactions that retard its movement.

The distribution of salts in the soil is also influenced by seedbed shape. Salts tend to accumulate to excess levels in certain regions of the seedbed under furrow irrigation. Seedbed and furrow shape can be designed to minimise this problem. Seed placement and surface irrigation strategies (e.g. alternative furrow, depth of water in furrows, etc.) that can also be used to optimise plant establishment under saline conditions are described by Kruse et al.. Sprinkler irrigation can be effective in leaching excessive salinity from the top-soil and in producing a favourable low-salinity environment in the upper soil layer which is necessary for the establishment of salt-sensitive seedings. However, other problems (such as foliar injury) are associated with sprinkling of saline water.

Saline, "bed-peaks" can be detopped to prevent exposure to emerging shoots. Under drip irrigation, the salt content is usually lowest in the soil immediately below and adjacent to the emitters and highest in the periphery of the wetted zone. Removal of salt that has accumulated in this wetting zone "front" must be addressed in the long-term. Susceptible crops should not be irrigated with saline water by sprinkler irrigation since their foliage absorbs salts upon wetting. Salts can accumulate in leaves by foliar absorption of such crops until lethal concentrations have been reached.

Crop sensitivity to saline sprinkling water is related more to the rate of foliar salt accumulation than to crop tolerance to soil salinity, per se. Hence, applications should be made during the night and in a manner to achieve frequent wetting ("washing") of the leaves in order to minimise foliar absorption of salts when irrigating with saline waters by sprinkler methods. The prevalent models of solute reactions and transport in irrigated soils suffer the deficiency of not appropriately representing the effects of the above-described processes that often occur under field conditions. Neither do they adequately account for the distribution uniformity effects of the irrigation application system itself, or of the infiltration uniformity effects resulting from variable soil permeability across the field.

Only recently has this problem been approached directly by measuring, on a large scale, solute distributions in field soil profiles. The results to date indicate that as yet no suitable method to quantify and integrate the effects of these processes on a field basis exists. It is probable that alternative modelling approaches, like that proposed by Corwin and Waggoner, may help in this regard. Some unique effects of irrigation are operative at the scale of whole projects and entire geohydrologic systems; hence, some management practices for salinity control should address this larger scale.

Some soil and water salination is inevitable with irrigation; the salt contained in the irrigation water remains in the soil as the pure water passes back to the atmosphere through the processes of evaporation and plant transpiration. Therefore, water in excess of evapotranspiration must be applied with irrigation to achieve leaching and prevent excess salt accumulation. This water must drain from the rootzone. Seepage from delivery canals also occurs in many irrigation projects. These drainage and seepage waters percolate through the underlying strata (often dissolving additional salts in the process), flow to lower elevation lands or waters and frequently cause problems there of waterlogging and salt-loading.

Saline soils typically are formed in such lands through the processes of evaporation. Ground and surface waters receiving these drainage and seepage waters typically are increased in salt concentration. The primary sources of return flow from an irrigation project are bypass water, canal seepage, deep percolation, and surface (tailwater) runoff. Bypass water is often required to maintain hydraulic head and adequate flow through a gravity-controlled canal system. It is usually returned directly to the river, and few pollutants, if any, are picked up in this route.

Evaporation losses from canals commonly amount to only a small percentage of the diverted water. Seepage from unlined canals is often substantial. It may contribute to high water tables, increase groundwater salinity and phreatophyte growth, and generally increases the amount and salinity of the required drainage from irrigated areas. Law et al. estimated that 20 percent of the total water diverted for irrigation in the USA is lost by seepage from conveyance and irrigation canals. If the water passes through salt-laden substrata or displaces saline groundwater, the salt pickup from this source can be substantial. Canal lining can reduce such salt loading.

Closed conduit conveyance systems can minimise both seepage and evaporation losses and ET by phreatophytes. The closed conduit system also provides the potential to increase project irrigation efficiency and to thus lower salt loading. Reducing the volume of water applied for irrigation proportionately reduces the amount of salt added and the amount needed to be removed by leaching. Minimising the leaching fraction maximises the precipitation of applied Ca, HCO_3, and SO_4 salts as carbonates and gypsum minerals in the soil, and it minimises the "pickup" of weathered and dissolved salts from the soil. The salt load from the rootzone can be reduced from about 2 to 12 tons/ha per year by reducing LF from 0.3 to 0.1.

Minimising leaching may or may not reduce salinity degradation of the receiving water where the drainage water is not intercepted and isolated and is returned to the associated surface or groundwater. A reduction of degradation will generally occur where saline groundwaters with concentrations in excess of those of the recharging rootzone drainage waters are displaced into the surface water or where additional salts, other than those derived from the irrigation water, are encountered in the drainage flow path and brought into solution by weathering and dissolution processes. Groundwaters receiving irrigation drainage water may not always benefit from reduced leaching.

With no sources of recharge other than drainage return flow, the groundwater eventually must come to the composition of the drainage water, which will be more saline with low leaching. Reduced leaching slows the arrival time of the leachate. Thus, the groundwater salinity may be lower with reduced leaching for an interim period of time. For groundwater under-saturated with $CaCO_3$ (unlikely in arid lands) being pumped for irrigation with no recharge other than by drainage return, groundwater will be slightly less saline under low leaching; groundwater saturated with $CaCO_3$ will show no benefit under low leaching; and groundwater saturated with $CaCO_3$ and nearing saturation with gypsum will show substantial benefit from low leaching.

Low leaching management can continuously reduce degradation of the groundwater, only if other sources of high-quality recharge into the basin exist and if flow out of the basin is high relative to drainage inflow. The extent to which leaching can be minimised is limited by the salt tolerances of the crops being grown, the irrigation system distribution uniformities and the variability in soil infiltration rates. In most irrigation projects, the currently used leaching fractions can be reduced appreciably without harming crops or soils, especially with improvements in irrigation management.

Management practices for the safe use of saline water for irrigation primarily consist of:

— selection of crops or crop varieties that will produce satisfactory yields under the existing or predicted conditions of salinity or sodicity;

— special planting procedures that minimise or compensate for salt accumulation in the vicinity of the seed;

— irrigation to maintain a relatively high level of soil moisture and to achieve periodic leaching of the soil;

— use of land preparation to increase the uniformity of water distribution and infiltration, leaching and removal of salinity;

— special treatments (such as tillage and additions of chemical amendments, organic matter and growing green manure crops) to maintain soil permeability and tilth. The crop grown, the quality of water used for irrigation, the rainfall pattern and climate, and the soil properties determine to a large degree the kind and extent of management practices needed.

Where salinity cannot be kept within acceptable limits by leaching, crops should be selected that can produce satisfactory yields under the resulting saline conditions. In selecting crops for saline soils, particular attention should be given to the salt tolerance of the crop during seedling development, because poor yields frequently result from failure to obtain a satisfactory stand. Some crops that are salt tolerant during later stages of growth are quite sensitive to salinity during early growth.

Failure to obtain a satisfactory stand of furrow-irrigated row crops on moderately saline soils is a serious problem in many places. This is because the rate of germination is reduced by excessive salinity. The failures are usually due to the accumulation of soluble salt in raised beds that are "wet-up" by furrow irrigation. Modifications in irrigation practice and bed shape should be used to reduce salt accumulation near the seed. The tendency of salts to accumulate near the seed during irrigation is greatest in single-row, round-topped beds (Figure 1). Sufficient salt to prevent germination may concentrate in the seed zone even if the average salt content of the soil is moderately low. Thus, such beds should be avoided when irrigating with saline waters using furrow methods, though "decapping" techniques may be used to advantage in this regard.

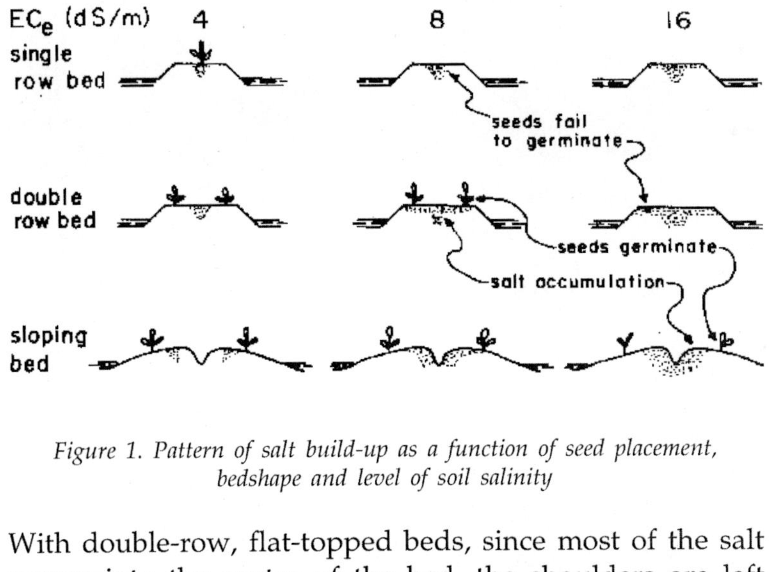

Figure 1. Pattern of salt build-up as a function of seed placement, bedshape and level of soil salinity

With double-row, flat-topped beds, since most of the salt moves into the centre of the bed, the shoulders are left relatively free of salt, thus seedling establishment may be enhanced by planting on the shoulders of such beds. Sloping beds are best for saline conditions because the seed can be safely planted on the slope below the zone of high salt accumulation. Such beds should be used, if possible, when furrow irrigating with saline waters. Planting in furrows or basins is satisfactory from the standpoint of salinity control but is often unfavourable for the emergence of many row crops because of problems related to crusting and poor aeration. This method is recommended only for the use of very saline irrigation waters and vigorous, hardy emerging plants.

Pre-emergence irrigation by use of sprinklers or special furrows placed close to the seed may be used to keep the soluble salt concentration low in the seedbed during germination and seedling establishment. After the seedlings are established, the special furrows may then be abandoned and new furrows made between the rows, and sprinkling replaced by furrow irrigation.

11

Purification of Streams and Rivers

Water pollution is a large set of adverse effects upon water bodies (lakes, rivers, oceans, groundwater) caused by human activities. Although natural phenomena such as volcanoes, storms, earthquakes etc. also cause major changes in water chemistry and the ecological status of water, these are not deemed to be pollution. Water pollution has many causes and characteristics. Humans and other farmed organisms produce bodily wastes which enter rivers, lakes, oceans and other surface waters. These wastes increase the solids suspended in the water (turbidity), they can increase the concentration of bacteria and viruses leading to potential health impacts.

Increases in nutrient loading may lead to eutrophication. Organic wastes such as sewage and farm waste impose high oxygen demands on the receiving water leading to oxygen depletion with potentially severe impacts on the whole ecosystem. Industries discharge a variety of pollutants in their wastewater including heavy metals, organic toxins, oils, nutrients, and solids. Discharges can also have thermal effects, especially those form power stations, and these too reduce the available oxygen.

Silt-bearing runoff from many activities including construction sites, forestry and farms can inhibit the penetration of sunlight through the water column restricting photosynthesis and causing blanketing of the lake or river bed which in turns damages the ecology. Pollutants in water include of a wide spectrum of chemicals, pathogens, and physical chemistry or sensory changes. Many of the chemical substances are toxic or even carcinogenic. Pathogens can obviously induce disease in either human or animal hosts. Alteration of water's physical chemistry include acidity, conductivity, temperature, and excessive nutrient loading (eutrophication). The impacts of agriculture on water quality are diverse.

Sources of Water Pollution

The most common sources of pollution of the streams and rivers. Any industry, in which water obtained from a water treatment system or a well comes in contact with a process or product, can add pollutants to the water. The resulting water is then classified as a wastewater. The wastewater can contain physical, chemical and biological pollutants in any form or quantity and cannot adequately be estimated without actual measuring and testing. The wastewater will either be discharged directly into a receiving body of water or into the sewerage system of a municipality, or they will be reused or recycled.

Table 1. Agricultural impacts on water quality

Agricultural activity	Impacts
	Surface water
Tillage/ploughing	Sediment/turbidity: sediments carry phosphorus and pesticides adsorbed to sediment particles; siltation of river beds and loss of habitat, spawning ground, etc.

Contd. . .

Fertilising	Runoff of nutrients, especially phosphorus, leading to eutrophication causing taste and odour in public water supply, excess algae growth leading to deoxygenation of water and fish kills.
Manure spreading	Carried out as a fertiliser activity; spreading on frozen ground results in high levels of contamination of receiving waters by pathogens, metals, phosphorus and nitrogen leading to eutrophication and potential contamination.
Feedlots/animal corrals	Contamination of surface water with many pathogens (bacteria, viruses, etc.) leading to chronic public health problems. Also contamination by metals contained in urine and faeces.
Irrigation	Runoff of salts leading to salinisation of surface waters; runoff of fertilisers and pesticides to surface waters with ecological damage, bioaccumulation in edible fish species, etc. High levels of trace elements such as selenium can occur with serious ecological damage and potential human health impacts.
Clear cutting	Erosion of land, leading to high levels of turbidity in rivers, siltation of bottom habitat, etc. Disruption and change of hydrologic regime, often with loss of perennial streams; causes public health problems due to loss of potable water.
Silviculture	Broad range of effects: pesticide runoff and contamination of surface water and fish; erosion and sedimentation problems.
Aquaculture	Release of pesticides (e.g. TBT (Tributyltin)) and high levels of nutrients to surface water and groundwater through feed and faeces, leading to serious eutrophication.

Areas unaffected by human activity can still pollute receiving streams due to stormwater runoff, which can be classified into animal, vegetable and soil sources. Stormwater transports industrial and municipal water pollutants to a receiving stream or an underground water supply. Public, private and industrial landfills can be a source of stormwater pollution because of runoff from the surface and underground leachate.

Point and Non-point Sources

From spatial perspective, the sources of pollution can be divided as point and non-point sources. As the name implies, point source pollution represents those activities where wastewater is routed directly into receiving water bodies by, for example, discharge pipes, where they can be easily measured and controlled. Examples of point sources include discharges from municipal wastewater treatment plants and industrial facilities and spills that occur due to accidents. Mass loading, for example, may be estimated by measuring the flow and solute concentration associated with a plant's effluent:

$$W = Q_{e.} C_e \tag{1}$$

where:

- W mass loading rate (M/T)
- Q_e volumetric flow rate for the point source
- C_e solute concentration in the effluent

In contrast, non-point source water pollution, once known as "diffuse" source pollution, arises from a broad group of human activities for which the pollutants have no obvious point of entry into receiving watercourses. An example of a non-point source is agricultural runoff that enters a stream as overland flow. During precipitation events, runoff from cultivated fields may contain pesticide residues and fertiliser, as well as suspended sediments that sorb contaminants. Loading due to this type of non-point source is represented by the lateral inflow terms in the general transport equation. Obviously, non-point source pollution is much more difficult to identify, measure and control than point sources.

Continuous and Instantaneous Sources

The sources of pollution and loading rates also vary with respect to time. Sources may be roughly classified according to their duration—continuous and instantaneous sources. Continuous sources introduce contaminants to the streams for extended periods of time. An example of a continuous source is the effluent from wastewater treatment plant. Although mass-loading rates may vary in time, many of these plants continuously discharge effluent into receiving waters, resulting in a continuous addition of solutes to the stream.

Instantaneous sources add contaminants to the streams over very short time periods. Although truly instantaneous sources do not exist, situation do arise in which contaminants are added to the stream over time intervals, which are short relative to the time-scales of interest. An example of this kind of source is an accidental spill, where contaminants enter a stream in a matter of minutes. This loading can be viewed as the instantaneous addition of mass at a discrete point in time.

Types of Pollutants

There are several ways of classifying pollutants in order to predict their effect on water quality and the means of their removal. First they can be classified as a solid, a liquid or gas, or as one of these mixed with or dissolved in another.

Physical Pollutants

Physical pollutants can be categorised as follows: solids content, solids type, colour, odour, taste, conductivity, temperature. Each of these categories can have chemical or biological sources.

Wastewater is "coloured" if it is not completely clear. Colour can be suspended colour (apparent colour) or dissolved colour (true colour). Suspended particles can be considered as suspended solids regardless of whether they are coloured. Dissolved colour may be caused by vegetable or mineral dyes, other inorganic industrial wastes, or by organic material from stormwater runoff. Dissolved colour should be considered as organic or inorganic chemical water quality and treaded as a chemical pollutant. Colour is quantitatively determined by visual comparison with a known concentration of coloured solutions or by spectrophotometric methods.

Odour like taste, is a measure of the effect of a stimulating substance on human respiratory membranes. Pure unpolluted water is a measure of zero odours. An odour in a wastewater is recognised as a potential environmental hazard. The accepted odour test is the Threshold Odour Test in which a sample is diluted with pure water until the least perceptible odour is noticed. The Minimum Detectable Threshold Odour Concentration (MDTOC) is reported as unit or dilutions required to reduce an odour to its detectable limit.

Taste is seldom listed as a permit requirement, but there are three methods of determination: flavour threshold test; flavour rating assessments; flavour profile analysis. The Flavour Threshold Test is similar to the odour test. The Flavour Rating Assessments is a scale for rating a drinking water as acceptable or not and Flavour Profile Analysis is a comparison between a wastewater taste and that of a documented sensory quality judged acceptable by trained testers. Physical pollutant recommendations are shown in Table 2 and Table 3 lists typical composition of domestic wastewater with no industrial wastes.

Table 2. Physical Pollutant Limitations

Pollutant	Public water supplies	
	Permissible	Desirable
Colour (units)	75	<10
Odour	Narrative	Virtually absent
Temperature	Narrative	Narrative
Turbidity	Narrative	Virtually absent

Table 3. Typical Domestic Wastewater Composition

Pollutant	Concentration, mg/l		
	Weak	Average	Strong
Total solids	350	800	1200
Total suspended solids	100	240	350
Total dissolved solids	250	500	850
Settleable solids (ml/l)	5	10	20
Volatile suspended solids	80	180	280
Volatile dissolved solids	100	260	300
Ammonia nitrogen	10	20	35
Total nitrogen	20	35	80
phosphorous	5	10	15
Alkalinity as $CaCO_3$	50	100	250
Oil, grease	50	100	150
5-day biochemical oxygen demand	120	225	400
Chemical oxygen demand	175	325	575
Total organic carbon	65	125	220

Chemical Pollutants

Chemical pollutants can be organic or inorganic. There are thousand of organic pollutants consisting of various combinations of carbon, hydrogen, and perhaps oxygen and/or many other inorganic or organic molecules. In general, organic pollutants are more biodegradable with fewer carbon and/or other molecules attached. Inorganic chemical pollutants can be categorised into pure chemical pollutants and chemical indicators tests (Figure 1).

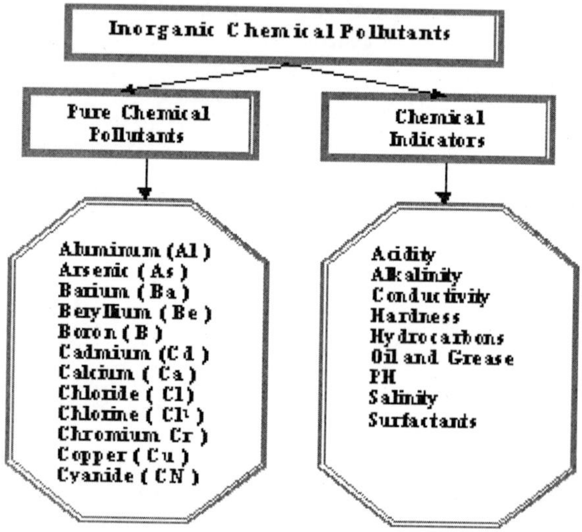

Figure 1. Classification of Inorganic chemical pollutants

Table 4 lists some of the pure chemical pollutants and their harmful effects.

Table 4. Pure chemical pollutants

Pure chemical pollutants	Harmful affect
Aluminium (Al)	is a contributor to Alzheimer's disease

Contd. . .

Arsenic (As)	is highly toxic to humans and is chronic, or cumulative in human and animal organs
Barium (Ba)	affects the gastrointestinal tract and the central nervous system
Beryllium (Be)	is toxic to fish and aquatic life and various plants and inhibits photosynthesis in terrestrial plants
Born (B)	at approximately 1 to 4 mg/l, can be toxic to plants and at approximately 30 mg/l, can have physiological effects on animals and humans.
Cadmium (Cd)	affects metabolism and is quite toxic to animals and humans; it is cumulative in the kidney and liver organs and can cause death.
Copper (Cu)	at levels above 100 mg/l is highly toxic to animals and humans and can cause vomiting and liver damage.
Cyanide (CN)	can be fatal at 8 mg/l. At low pH, CN forms hydrogen cyanide (HCN), which is a highly toxic, almond smelling mustard gas.
Lead (Pb)	it is cumulative in animal and human organs; with concentration greater than 0,5 mg/l causes lead poisoning. Plumbism is a disease caused by lead that affects the central nervous system of animals and humans.

Certain tests have been developed to indicate water quality based on chemical characteristics that can be simpler, less expensive or more indicative of water quality than a chemical compounds tests. $CaCO_3$ is used as a standard for many of the indicator tests since its molecular weight is 100 and calculations are simplified.

Biological Pollutants

As biological pollutants use oxygen for energy, the tests

can involve the measurement of either carbon content or oxygen demand, or an actual bacterial count. The Biological Oxygen Demand (BOD) test is used in order to determine the oxygen requirements of wastewaters. The BOD, unless treated with an ammonia-inhibiting chemical, will indicate the total of carbonaceous and nitrogenous oxygen demand. A 250 to 300 ml airtight bottle is filled with sample to overflowing. Dissolved oxygen is measured initially after dilution and after incubation at 20°C for the time period of the test. It is necessary to have a population of microorganisms present that is capable of degrading the organic material in the sample.

Many unchlorinated domestic or industrial wastewaters will afford sufficient bacteria for this purpose. This population is called "the seed". One BOD test must be run on a bottle containing seed and dilution water. The dilutions should be made so that the least dilution results in at least a 50% dissolved oxygen depletion. The seed is also added and diluted so that the residual dissolved oxygen is at least 1 mg/l and the uptake at least 2 mg/l after 5 days. Several dilutions of the prepared sample should be made to ascertain that these results would be obtained.

Chemical Oxygen Demand (COD) test is a measure of the oxygen required from a strong chemical oxidant for the destruction of an organic material. The chemical oxidant selected for the COD test is potassium dichromate ($K_2Cr_2O_7$). The sample is oxidised by a boiling mixture of $K_2Cr_2O_7$ and sulfuric acid using an excess of $K_2Cr_2O_7$. After digestion, the remaining unreduced $K_2Cr_2O_7$ is measured to determine the amount consumed by titration with ferrous ammonia sulphate (FAS) until the end point is indicated by sharp colour change. The COD is calculated as:

$$COD = 8000 \cdot M \cdot (A\text{-}B)/\text{ml sample} \qquad (2)$$

where:

COD	Chemical Oxygen Demand
A	ml FAS used for blank
B	ml FAS used for sample;
M	molarity of FAS.

The blank is a test with all reagents without the sample. The COD takes about 2 to 3 hours and can be run with a relatively simple and inexpensive kit.

Total Organic Carbon (TOC) test is more direct indication of organic content than the BOD or COD test. The TOC test indicates the total organic matter present and is independent of the oxidation state of the pollutant. This test converts and removes inorganic carbon (CO + CO_2) and measures total organic carbon by converting it into CO_2 in a heated reaction chamber. The CO_2 is measured using infrared methods. Theoretical Oxygen Demand (THOD) is the stoichiometrically determined oxygen needed to convert all carbon molecules in pollutants to CO_2, and all NH_3 by balancing equations. This calculation is possible on certain rather pure industrial waste, but is impractical for most wastewaters, especially those containing domestic sewage or containing vegetable or animal wastes.

Relationship between BOD, COD, TOC and ThOD. After biological oxidation, approximately ten percent of the original organics (completely biodegradable wastewater) remain as non-biodegradable cellular residues. Hence, the cellular residues are not measured by the BOD test. Therefore, depending on the type of wastewater:

$$BoDu = 0{,}9ThOD \qquad (3)$$

where:

BoDu ultimate BOD

ThOD theoretical oxygen demand

For domestic sewage and some biodegradable industrial wastes, the relationship between BOD_5 and BODu is:

$$BOD_5 = 0.77BODu \qquad (4)$$

where:

BOD_5 = 5 day BOD

BoDu ultimate BOD

For most wastewaters: ThOD = COD. Where ThOD is the theoretical oxygen demand; COD is the chemical oxygen demand since the COD test oxidises all organics except those that are totally resistant to dichromate oxidation.

Stoichiometrically, the COD/TOC ratio should be approximately the molecular ratio of oxygen to carbon:

$$COD/TOC = 32/12 = 2.66 \qquad (5)$$

The ratio will actually range from zero, when organic material is resistant to dichromate oxidation, to as much as 6.0 when inorganic reducing agents are present. For raw domestic sewage and some biodegradable industrial wastes, the following ratio occurs:

$$BOD_5/TOC = 32/12 \cdot (0.90) \cdot (0.77) = 1.85 \qquad (6)$$

where:

BOD_5 = 5 day BOD

TOC = Total Organic Carbon

$$0{,}90 = BODu/ThOD \quad (7)$$

as

$$0{,}77 = BOD_5/BODu \quad (8)$$

Conventional and Toxic Pollutants

The major differences between conventional and toxic pollutants relate to four areas:

— *Natural vs. alien*: The conventional pollutant typically deals with the natural cycle of organic production and decomposition. The pollution problem involves an overstimulation of the natural processes governing the waterbody. In contrast, many toxins do not occur naturally. Examples are pesticides and other synthetic organics. For such alien pollutants, the problem is one of poisoning or interference with natural processes.

— *Aesthetics vs. health*: Conventional pollutants deal solely with aesthetics, in contrast, the toxic substances issue is almost totally dominated by health concerns. Most toxic remediations focus on the contamination of drinking water and an aquatic food stuffs.

— *Few vs. many*: Conventional waterquality management deals with on the order of about 10 pollutants. In contrast, there are thousands of organic chemicals that could potentially be introduced into our natural waters.

— *Single species vs. solid/liquid partitioning*: In most cases, conventional pollutants are treated as a

> single species. Consequently, its strength in the water body is measured by a single concentration—for example, dissolved oxygen deficit. In contrast, the transport, fate and ecosystem impact of toxicants is intimately connected with how they partition or associate with solid matter in and below the water body. Thus toxic substance analysis must distinguish between dissolved and particulate forms. The distinction between dissolved and particle-associated toxicant has an impact on transport and fate in the sense that certain mechanisms differentially impact the two forms.

The purpose of a toxicity test is to determine the concentration of a pollutant, which can exist in a body of water, that will neither kill a test organism (acute toxicity) nor prevent reproduction (chronic toxicity) of the organism. After this determination an application factor is usually multiplied by the test results to obtain a permit limit. Commonly used application factors are 0.01 if the pollutions are cumulative in the internal organs, and 0.10 if the pollutant is not cumulative. The most common toxicity tests are for daphnia and fish. Other standard tests are toxicity to algae, phytoplankton, protozoa, coral annelids, molluscs, acartia tonsa and aquatic insects.

Classification Based on Sources

Pollutants from Urban Areas

The pollutants that occur in urban areas vary widely, from common organic material to highly toxic metals. Some pollutants, such as insecticides, road salts and fertilisers, are intentionally placed in the urban environment. Other pollutants, including lead from automobile exhaust and oil drippings from trucks and cars, are the indirect result of urban activities.

Agricultural Pollutants

Agriculture is only one among various causes of non-point sources of pollution, however it is generally regarded as the largest contributor of pollutants of all the categories. Pollutants, irrespective of source, are transported overland and through the soil by rainwater and melting snow. These pollutants ultimately find their way into groundwater, wetlands, rivers and lakes and, finally, to oceans in the form of sediment and chemical loads carried by rivers. The ecological impact of these pollutants range from simple nuisance substances to severe ecological impacts involving fish, birds and mammals, and on human health.

Sediment, nutrients and pesticides occupy the first four categories in Table 1 and are significantly associated with agriculture. Although agriculture contributes to a wide range of water quality problems, anthropogenic erosion and sedimentation is a global issue that tends to be primarily associated with agriculture. While there are no global figures, it is probable that agriculture, in the broadest context, is responsible for much of the global sediment supply to rivers, lakes, estuaries and finally into the world's oceans. Pollution by sediment has two major dimensions—physical and chemical dimensions.

- — Physical dimension—top soil loss and land degradation by gullying and sheet erosion; leads both to excessive levels of turbidity in receiving waters, and to off-site ecological and physical impacts from deposition in river and lake beds.
- — Chemical dimension—the silt and clay fraction (<63 μm fraction), is a primary carrier of adsorbed

chemicals, especially phosphorus, chlorinated pesticides and most metals, which are transported by sediment into the aquatic system.

Sediment, as a physical pollutant, impacts receiving waters in the following principal ways:

— High levels of turbidity limits penetration of sunlight into the water column, thereby limiting or prohibiting growth of algae and rooted aquatic plants. In spawning rivers, gravel beds are blanketed with fine sediment which inhibits or prevents spawning of fish. In either case, the consequence is disruption of the aquatic ecosystem by destruction of habitat. Notwithstanding these undesirable effects, the hypertrophic (nutrient rich) status of many shallow lakes, especially in developing countries, would give rise to immense growth of algae and rooted plants were it not for the limiting effect of light extinction due to high turbidity. In this sense, high turbidity can be "beneficial" in highly eutrophic lakes; nevertheless, many countries recognise that this situation is undesirable for both aesthetic and economic reasons and are seeking means to reduce both turbidity and nutrient levels.

— High levels of sedimentation in rivers leads to physical disruption of the hydraulic characteristics of the channel. This can have serious impacts on navigation through reduction in depth of the channel, and can lead to increased flooding because of reductions in capacity of the river channel to efficiently route water through the drainage basin. For example, calculations by the UFRGS of erosion and sediment transport in the Sao Francisco River Basin, a large drainage system in eastern Brazil, demonstrate that the central portion of the river basin is now dominated by sediment deposition. This has resulted in serious disruption of river transportation, and clogs hydraulic facilities that have been built to provide irrigation water from the main river channel. The sediments largely originate from rapidly eroding subbasins due to poor agricultural practices.

The role of sediment as a chemical pollutant is a function of the chemical load that is carried by sediments. The role of sediment in chemical pollution is tied both to the particle size of sediment, and to the amount of particulate organic carbon associated with the sediment. The chemically active fraction of sediment is usually cited as that portion which is smaller than 63 μm (silt + clay) fraction. For phosphorus and metals, particle size is of primary importance due to the large surface area of very small particles. Phosphorus and metals tend to be highly attracted to ionic exchange sites that are associated with clay particles and with the iron and manganese coatings that commonly occur on these small particles.

Many of the persistent, bioaccumulating and toxic organic contaminants, especially chlorinated compounds including many pesticides, are strongly associated with sediment and especially with the organic carbon that is transported as part of the sediment load in rivers. Measurement of phosphorus transport in North America and Europe indicate that as much as 90% of the total phosphorus flux in rivers can be in association with suspended sediment. The affinity for particulate matter by an organic chemical is described by its octanol-water partitioning coefficient (KOW). This partitioning coefficient is well known for most organic chemicals and is the basis for predicting the environmental fate of organic chemicals.

Chemicals with low values of KOW are readily soluble, whereas those with high values of KOW are described as "hydrophobic" and tend to be associated with particulates. Chlorinated compounds such as DDT and other chlorinated pesticides are very hydrophobic and are not, therefore, easily analysed in water samples due to the very low solubility of the chemical. For organic chemicals, the most important component of the sediment load appears to be the particulate organic carbon fraction that is transported as part of the sediment. Scientists have further refined the partitioning coefficient to describe the association with the organic carbon fraction (KOC).

Another important variable is the concentration of sediment, especially the <63 μm fraction, in the water column. Even those chemicals that are highly hydrophobic will be found in trace levels in soluble form. Where the suspended load is very small (less than 25 mg/l), the amount of water is so large relative to the amount of sediment that the bulk of the load of the chemical may be in the soluble fraction. This becomes an important issue in the monitoring of hydrophobic chemicals. Unlike phosphorus and metals, the transport and fate of sediment-associated organic chemicals is complicated by microbial degradation that occurs during sediment transport in rivers and in deposited sediment.

Nevertheless, the role of sediment in the transport and fate of agricultural chemicals, both for nutrients, metals, and pesticides is well known and must be taken into account when monitoring for these chemicals, and when applying models as a means of determining optimal management strategies at the field and watershed level. Fertilisers cause the following problems FAO/ECE:

— Fertilisation of surface waters (eutrophication) results in, for example, explosive growth of algae that causes disruptive changes to the biological equilibrium (including fish kills). This is true both for inland waters (ditches, river, lakes) and coastal waters.

— Groundwater is being polluted mainly by nitrates. In all countries groundwater is an important source of drinking water. In several areas the groundwater is polluted to an extent that it is no longer fit to be used as drinking water according to present standards.

While these problems were primarily attributed to mineral fertilisers by FAO/ECE, in some areas the problem is particularly associated with extensive and intensive application of organic fertilisers (manure). The importance and, in some cases, the major problems associated with organic fertilisers, deserve special mention. Manure produced by cattle, pigs and poultry are used as organic fertiliser the world over. To this is added human excreta, especially in some Asian countries where animal and human excreta are traditionally used in fish culture as well as on soils.

However, intensive livestock production has produced major problems of environmental degradation, a phenomenon that has been the subject of European and North American legislation and control. The problem is particularly acute in areas of intensive livestock production, such as in the Eastern and Southern parts of the Netherlands where the production of manure greatly exceeds the capacity of the land to assimilate these wastes. In addition to problems associated with excessive application of manure on the land, is the problem of direct runoff from intensive cattle, pig and poultry farms. Although this is controlled in many western countries, it constitutes a serious problem for water quality in much of the rest of the world.

To the typical pathways of degradation, that of surface runoff and infiltration into the groundwater, is added the volatilisation of ammonia, which adds to acidification of land and water. In a review of environmental impacts caused by animal husbandry in Europe, the FAO/ECE reported the following major categories of impacts:

— Fertilisation of surface waters, both as a result of direct discharges of manure and as a consequence of nitrate, phosphate and potassium being leached from the soil.

— Contamination of the groundwater as a result of leaching, especially by nitrate. Phosphates are less readily leached out, but in areas where the soil is saturated with phosphate this substance is found in the groundwater more and more often.

— Surface waters and the groundwater are being contaminated by heavy metals. High concentrations of these substances pose a threat to the health of man and animals. To a certain extent these heavy metals accumulate in the soil, from which they are taken up by crops. For example, pig manure contains significant quantities of copper.

— Acidification as a result of ammonia emission (volatilisation) from livestock accommodation, manure storage facilities, and manure being spread on the land. Ammonia constitutes a major contribution to the acidification of the environment, especially in areas with considerable intensive livestock farming.

The term "pesticide" is a composite term that includes all chemicals that are used to kill or control pests. In agriculture, this includes herbicides (weeds), insecticides

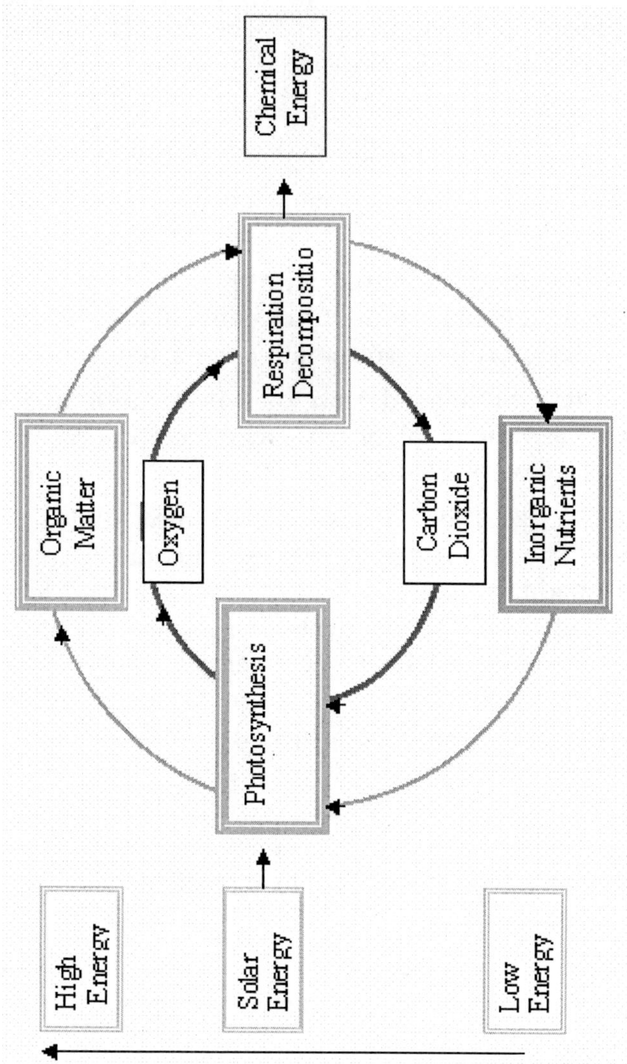

Figure 2. The natural cycle of organic production and decomposition

(insects), fungicides (fungi), nematocides (nematodes), and rodenticides (vertebrate poisons). Agricultural use of pesticides is a subset of the larger spectrum of industrial chemicals used in modern society. The impact on water quality by pesticides is associated with the following factors:

- — Active ingredient in the pesticide formulation.
- — Contaminants that exist as impurities in the active ingredient.
- — Additives that are mixed with the active ingredient (wetting agents, diluents or solvents, extenders, adhesives, buffers, preservatives and emulsifiers).
- — Products of degradation that is formed during chemical, microbial or photochemical degradation of the active ingredient.

In addition to use of pesticides in agriculture, silviculture also makes extensive use of pesticides. In some countries, such as Canada, where one in ten jobs is in the forest industry, control of forest pests, especially insects, is considered by the industry to be essential. Aircraft often spray insecticides over very large areas. Irrigated agriculture, especially in tropical and subtropical environments, usually requires modification of the hydrological regime which, in turn, creates habitat that is conducive to breeding of insects such as mosquitoes which are responsible for a variety of vector-borne diseases.

In addition to pesticides used in the normal course of irrigated agriculture, control of vector-borne diseases may require additional application of insecticides such as DDT, which have serious and widespread ecological consequences. In order to address this problem, environmental management methods to control breeding

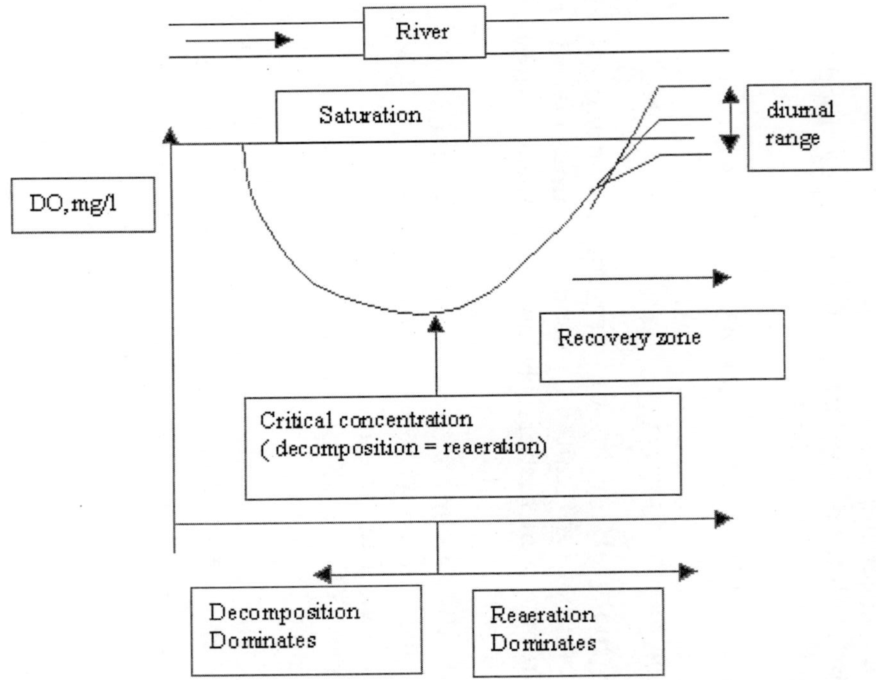

Figure 3. The dissolved oxygen "sag" that occurs below sewage discharges into streams

of disease vectors are being developed and tested in many irrigation projects.

Pollutants from Municipal Sources

The pollutants from non-industrial municipal sources are all biological and as such can be readily biodegraded. The main pollutants are: feces, urine, paper, food waste, laundry wastewater and etc.

Impacts of Pollutants on the Water Quality

Stream Dissolved Oxygen

The natural cycle of organic production and decomposition can be used to understand the environment in a stream below a wastewater discharge (Figure 2).

If the river is unpolluted, dissolved oxygen concentrations above the discharge will be near saturation. The introduction of the sewage will elevate the levels of both dissolved and solid organic matter. This has three impacts:

- The solid matter makes the water turbid and unsightly. Thus light cannot penetrate and plant growth is suppressed;
- The solids settle downstream from the sewage outfall and create sludge beds that can emit noxious odours;
- The organic matter provides food for heterotrophic organisms. Large populations of decomposer organisms break down the organic matter in the water and in the process deplete the dissolved oxygen. In addition, breakdown of organic matter takes place in the sludge bed, and a sediment

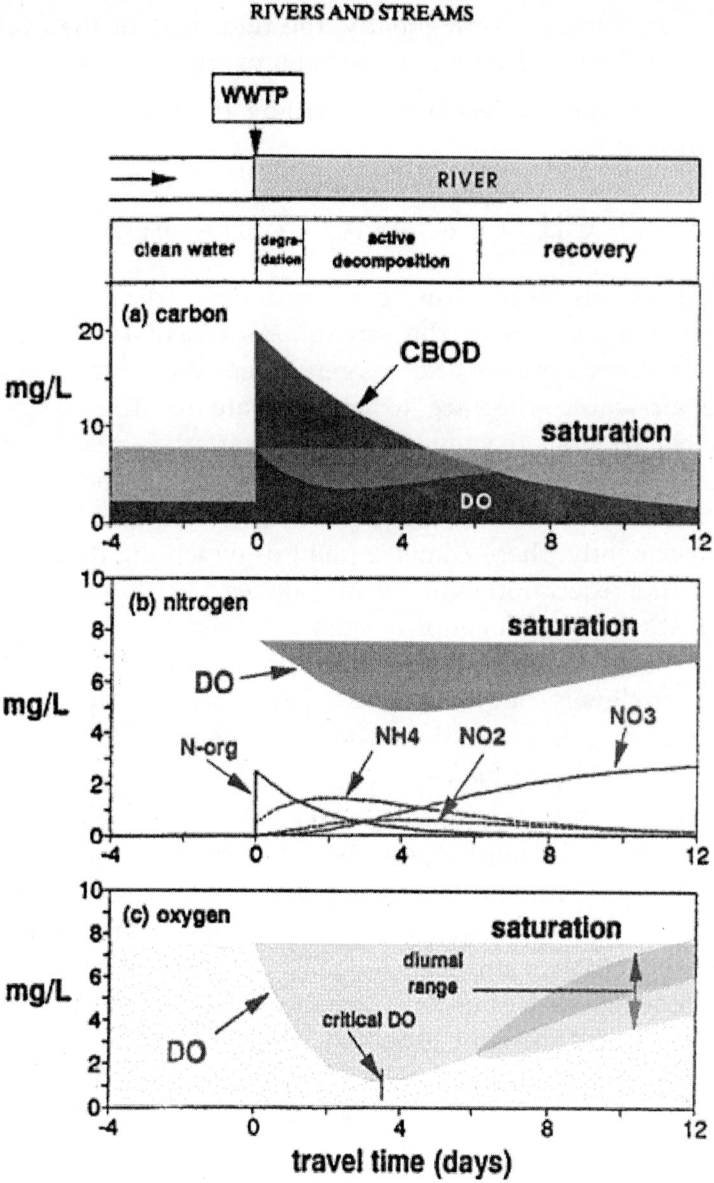

Figure 4. Trends of (a) carbon, (b) nitrogen, and (c) oxygen below a wastewater treatment plant discharge into a river

oxygen demand supplements oxygen depletion in the water. Consequently, the right side of the cycle in Figure 2 begins to become overemphasised.

The fundamental reactions governing the process can be expressed by:

$$C_6H_{12}O_6 + 6O_2 \rightarrow 6CO_2 + 6H_2 + 6H_2O \qquad (9)$$

Aside from depletion, counteracting force acts to replenish oxygen in the stream. As dissolved oxygen levels drop, atmospheric oxygen enters the water across the air-water interface to compensate for the oxygen deficit. At first, oxygen depletion dwarfs this reaeration. However, as the organic matter is assimilated, the rate of depletion diminishes as the reaeration rate increases. Consequently, there comes a point at which the depletion and the reaeration will be in balance. At this point, a lowest or critical level of oxygen will be reached. Beyond this critical level, the reaeration process dominates and oxygen levels begin to rise again. Thus, a dissolved oxygen sag is created in the river below the sewage discharge (Figure 3).

Along with decomposition, the water becomes clearer as much of the solid matter from the discharge settles to the bottom. In addition, inorganic nutrients liberated during the decomposition process are high. Consequently, the reaction 9 reverses and plant photosynthesis begins to dominate in the recovery zone. As the result of plant growth, carbon dioxide and water are utilised and oxygen is produced. Consequently, large diurnal (that is over the daily cycle) oxygen swings can occur in the recovery zone.

Bacteria and Other Pathogens

Beyond the chemical changes, sewage also leads to

significant effects on the biota. The bacterial populations flourish until the BOD and/or the oxygen is exhausted. In addition, the bacteria themselves provide a food source for a succession of organisms, consisting of ciliates, rotifers, and crustaceans. The diversity of higher organisms decreases drastically in the degradation and active decomposition zones below the discharge. At the same time, total number of organisms increases. As recovery ensues, these trends are reversed.

Nitrogen Compounds

Nitrification

Nitrogen compounds in wastewater also have an impact on a river's oxygen resources. Sewage nitrogen can be broadly broken down into organic nitrogen compounds (such as proteins and urea) and ammonia. With time the organic nitrogen compounds are hydrolysed to create additional ammonia. Autotrophic bacteria then assimilate the ammonia and create nitrite (NO_2^-) and nitrate (NO_3^-). The conversion of ammonia to nitrate is called nitrification. Along with the decomposition of carbonaceous matter, nitrification will cause a dissolved oxygen sag in streams and rivers. The entire process, as depicted in Figure 4, shows the collective effect of organic carbon and nitrogen decomposition on a stream's oxygen recourses.

The carbonaceous BOD is elevated at the point where the sewage enters the stream. It then decreases as the CBOD is decomposed and settles. The effect of decomposition in tandem with reaeration results in an oxygen sag (Figure 4.a). The following Figure 4.b, nitrification also induces a sag. However, because of the sequential nature of the nitrification reactions, the minimum occurs farther downstream than for the carbon induced depletion. Finally, Figure 4.c shows the total depletion that results from the combined effect of carbon and nitrogen.

The resulting DO profile is worse than the individual sag from carbon and nitrogen alone, and critical DO level occurs at a distance somewhere between the carbon and the nitrogen cases. The ultimate result of the nitrification process is nitrate. In sufficiently high concentrations, nitrate in drinking water can cause serious and occasionally fatal effects to infants.

Ammonia toxicity

Ammonia exists in two forms in natural water—ammonia ion (NH_4) and ammonia gas (NH_3). The un-ionised form is toxic to fish. For example, at moderate temperatures (20°C) and pH levels above 9, upwards of 20 percent of the total ammonia will be in the un-ionised form. The most noteworthy instance occurs in the recovery zone below a secondary sewage discharge into a shallow stream. For such cases, suspended solids will be low and ammonia will be high. Thus, profuse plant growth can occur. In the late afternoon, such photosynthesis can deplete carbon dioxide from the water and, for weakly buffered waters, induce a large increase in pH. The late afternoon is also coincidentally the time of day when temperatures are highest. Consequently, the high pH and temperature can result in elevated levels of un-ionised ammonia.

Eutrophication

Apart from its other characteristics, nitrogen serves as an essential nutrient for plant growth. There are two aspects of nitrogen as a nutrient that bear mentioning. First, algae use both ammonium and nitrate as a source of nutrient. They prefer ammonium, but are capable of generating an enzyme that allows them to utilise nitrate when ammonium levels are low. Second certain blue-green algae are capable of fixing free nitrogen. This gives them a competitive advantage in situations where other forms of nitrogen are in short supply. In general, stream eutrophication can have a number of deleterious effects on a stream or river.

— The profuse growth of plants decreases water clarity and some species from unsightly scums.

— Certain species of algae cause taste and odour problems in drinking water.

— Certain blue-green algae can be toxic when consumed by animals.

— Eutrophication can alter the species composition of a river ecosystem. Native biota may be displaced as the environment becomes more productive.

— The nutrients can indirectly affect other aspects of stream chemistry. For example, the uptake and release of carbon dioxide by plants can alter the system's pH.

Careful grading of land makes possible a more uniform application of water and, hence, better salinity control when irrigating with saline water. Barren or poor areas in otherwise productive fields are often either high spots that do not receive enough water for good crop growth or for leaching purposes or low spots that remain too wet for seedling establishment. Lands that have been irrigated one or two years after initial grading usually need to be regraded to remove the surface unevenness caused by the settling of fill material. Annual crops should be grown after the first grading so that regrading can be performed before a perennial crop is planted. Land levelling causes a significant soil compaction due to the weight of the heavy equipment and it is advisable to follow this operation with subsoiling, chiselling and ploughing to break up the compaction and restore or improve water infiltration.

12

Long-distance Water Transfer

Water is a renewable resource and, unlike non-renewable resources like oil or natural gas, there is no danger that the world is going to run out of water. The principal problem with water is that its distribution varies tremendously both with respect to time and space, and accordingly its rational management is of paramount importance for the welfare of mankind. The extent of water use for any one purpose varies from one country to another, and is dependent on a variety of factors like state of economic development, including standard of living; importance and extent of a specific sector like industry or agriculture in the national economy; efficiency of water use; socio-cultural practices, and so on. The largest user of water and accounts for nearly 80 per cent of total consumption.

With increasing global population, more and more water will be required for different uses. Furthermore, current evidence indicates that water requirement per capita increases with improvements in the standard of living. It is, however, extremely difficult to make meaningful projections for future global water requirements since efficiency of use is a most important consideration.

Two major water users are the agricultural and industrial sectors, and the potential for more efficient water management in both these sectors is enormous. For example, identical industrial products are being manufactured in the same country, say the United States, where one factory uses only 2 per cent of the total water needs of another, per unit of product, by using extensive in-plant recirculation and treatment technologies. There is no doubt that the agricultural sector is the major user of water for most developing countries. Water requirements naturally vary tremendously from one country to another, depending on different factors, among which are total land under agricultural production, types of soils and crops cultivated, climate, availability and pricing of irrigation water, efficiency of water use and management practices.

Agriculture is water intensive and, on average, it takes approximately 1,000 tons of water to grow one ton of grain and 2,000 tons of water to grow one ton of rice. In addition, both animal husbandry and fisheries require abundant water. If the current trends continue, future water requirements are going to increase tremendously. In 1975, the total area irrigated in the world amounted to 223 million ha, of which 93 million ha was in developing countries.

Some 15 per cent of the world's cropland is irrigated, but it yields from 30 to 40 per cent of all agricultural production. The amount of water used by irrigated crops is nearly 1,300,000 million m^3 but, because of losses in storage, conveyance and use, the total amount used increases to almost 3,000,000 million m^3. It is estimated that by the end of the present decade in 1990, the total area irrigated in the world will increase to 273 million ha, of which 119 million ha will be in developing countries.

Without major improvements in efficiency for water use, more and more water will be required to sustain and improve present agricultural and industrial developments and more will be necessary for future developments. Certain parts of the world have already started to face water shortages, and if the present trends continue, the situation is likely to become worse in the future. Some people have already claimed that water will soon become one of the major constraints for future economic developments.

Groundwater is being increasingly exploited for agricultural development. In many parts of the world, groundwater withdrawal rates are exceeding natural recharge rates, and thus contributing to groundwater mining. Even in an advanced country like the United States, groundwater use is rising more rapidly than surface water use, and nearly 25 per cent of all groundwater withdrawals, in 1975 in that country amounted to mining. For example, in the San Joaquin Basin in California, United States, approximately $1.85x10^9$ m^3 more water is pumped every year than is recharged naturally. Since this overdraft accounts for some 12.5 per cent of San Joaquin's annual water supply, its long-term sustainability has to be seriously questioned. The situation is somewhat similar or even worse in most arid developing countries.

In nearly all arid developing countries, faced with the twin problems of increasing population and the need to feed that population properly, planners and decision-makers are looking for means to increase irrigation water availability so as to increase agricultural production. They are exploring alternatives to alleviate what often could be considered rapidly developing critical situations. One possibility several countries are exploring at present is to transfer water over long distances from water-surplus areas to deficient areas. Already, some projects have been constructed in different parts of the world to divert water from one basin to another. Long-distance mass transfer of water is, however, only one of several alternatives of non-conventional water development.

Several other possibilities are being explored, among which are weather modification, desalination, iceberg towing, and the use of very large crude carriers (VLCC) to transport fresh water to water-deficient areas. Some of these alternatives are still in the research and development stage, and cannot be used economically for large scale agricultural development. Others, like desalination, can be used only for site-specific solutions, and can produce only a limited quantity of water, which means it is somewhat unlikely at present to be used economically for extensive agricultural development.

Long Distance Water Transfer

Conceptually, it can be argued that all water development projects involve transfer of water over long distances. Part of rain or snow that falls in one region ultimately finds its way, as surface and groundwater, to a river, which often flows through several regions. At certain locations, water can be stored in reservoirs by damming the river. The water stored in reservoirs can be released as and when required. In other words, water is always in motion and is being continually transferred from one region to another by both natural and artificial means.

Before long-distance mass transfer of water can be seriously investigated as a viable policy option, it is essential to carry out a comprehensive assessment of available water resources-both surface and groundwater-in terms of quantity as well as quality. It is a fundamental fact of hydrology that availability of water in a river is a function of both space and time. Thus, static assessment is not a viable option: time-series data of hydrological variables- either real or reconstructed-is an important prerequisite for any analysis. In order that reliable forecasts of water availability can be made, it is necessary to have adequate data over a reasonable period of time. Based on such data, long-term water management plans can be prepared.

However, in many parts of the world, such data are not available or, if available, they are for a rather limited time period. Furthermore, very rarely is any information available on their reliability and accuracy. This is unfortunately the case for most developing countries. This is basically the situation with data on water quantity; information on water quality is even more scarce and seldom available, especially in developing countries. Quality of water determines its possible uses, and hence data on water quality are necessary to develop rational management plans. Much progress has to be made in this field.

Other problems associated with data on water quantity and quality, in addition to data scarcity, are non-representativeness of monitoring sites, unavailability of trained technicians to collect data and maintain data collection instruments and monitoring networks, and lack of experienced professionals to analyse the data collected and facilities to store and retrieve data. The situation is further complicated by the presence of different agencies at federal, state and local levels who collect water-related data. Co-operation between these agencies is generally poor and sometimes even non-existent.

Accordingly, exchange of data is not always feasible. In the United States alone, at least 20 federal agencies are associated solely with water quality monitoring programmes. Proliferation of data collection agencies often makes the system complicated and in a few cases even unmanageable. The second important consideration is the assessment of water demands for different purposes. It should be noted that the term "demand", in the context of water resources management, generally means requirements, and is very rarely used in its traditional economic sense. Indeed, very rarely is the concept of demand elasticity explicitly considered within the water planning process per se.

Consequently very little is known about constructing realistic demand functions under varying socio-economic considerations. Expressed differently, it means that emphasis so far has been on supply management-that is, increase in supply is considered to be virtually the only management alternative-rather than consideration of demand management. As the water requirements for various purposes continue to increase and available sources become more and more exploited and polluted, it is highly likely that emphasis will gradually shift from supply to demand management.

Efficiency of existing water use is the third important consideration. Undoubtedly, the agricultural sector is one of the most inefficient users of water, and significant improvements could be made here. Existing efficiencies of irrigation systems are so low that they do not by any means reflect the actual water requirements of crops being irrigated. As mentioned, on a global basis, 1,300,000 million m^3 of water is used for irrigation crops, but for this 3,000,000 million m^3 of water have to be withdrawn. In other words 57 per cent of total water withdrawn is lost in the process. This, however, should not be construed to mean that the remaining 43 per cent is efficiently used.

Over-irrigation is endemic and not exactly an uncommon practice in both developed and developing countries. It not only means that water is wastefully used but also contributes to development of environmental problems, like increase in groundwater table and salinity level.

One of the most inefficient parts of the irrigation system is often the section where water is transferred from canal outlets to farms. It has become a no-man's land due to undefined responsibility, which in turn contributes to improper design and unsatisfactory operation and maintenance. While much research has been carried out on losses from the canal itself, very little work has been done on losses from the transfer sections. Studies carried out on 40 such sections in the Indus Basin during 1975 and 1976 indicated losses ranging from 33 to 65 per cent, with an average of 47 per cent.

Another investigation on 60 sections carried out in 1977 and 1978 by Water and Power Development Authority of Pakistan indicated similar losses. The magnitude of this problem can best be realised by considering the case of good lined canals, which are expensive to construct and have operating efficiencies of 70 to 80 per cent. When the efficiency of the total system is considered, i.e. lined canals in conjunction with the inefficient section from canal outlets to farms, the total efficiency is of the order of 20 to 50 per cent, which means that even for expensive, lined and well-maintained canal systems, in many cases only one fifth of water released from a reservoir reaches the crops being irrigated. A major result of this sad state of affairs is that engineers and planners have accepted this inefficient system, at least implicitly.

During planning of irrigation projects, total water requirements are generally calculated by multiplying the area to be irrigated by the water required per hectare. The water requirement per hectare is often estimated on the basis of existing systems, where major portions of water released from reservoirs are lost. Accordingly, overall estimates of irrigation water requirements are invariably high-certainly much higher than necessary-and the inefficient system is condoned and perpetuated. In other words, most irrigation systems designed so far are inefficient and use far more water than necessary.

Unfortunately, instead of attempting to make irrigation systems more efficient and then maintaining them at such levels, engineers are constantly seeking new sources of water for irrigation. They often look for costly alternatives, like interbasin water transfer, when such major and expensive projects are not essential and cheaper alternatives are available-which can be implemented within a much shorter time frame with indigenous labour and expertise-by simply improving the existing systems. Furthermore, even if new projects are developed, unless special efforts are made to maintain their efficiencies at high levels, their effectiveness will decline with time and thus the vicious circle will continue to be perpetuated.

Several proposals have been put forward in North America in recent years for long-distance mass transfer of water, but undoubtedly the most ambitious one suggested so far is the North American Water and Power Alliance (NAWPA). First suggested in 1964 by Ralph M. Parsons Company of Los Angeles, this $10billion-plus project kindled the interest of many engineers and politicians. Within some six years of the NAWPA scheme being proposed, a whole series of other schemes were put forward to re-distribute the water resources availability in North America. Most of these proposals cannot be considered very seriously at present, since some were based literally on drawing lines on topograpical maps, with very little, if any, analyses of their engineering, economic, political and social feasibilities.

The NAWPA project will be briefly outlined here primarily for two reasons: (a) its gigantic scale; and (b) more technical information is available on this scheme than on others. The fundamental concept of NAWPA is to collect surplus water from the high precipitation areas of the north-western part of the North American continent and distribute it to water-deficient areas in Canada, the United States and northern Mexico. A series of dams and power stations in Alaska and northern British Columbia would collect water and provide power to pump this water up to the Rocky Mountain Trench Reservoir in south-eastern British Columbia.

From the Rocky Mountain Trench Reservoir, water would be lifted by pumps to the Sawtooth Reservoir in Central Idaho. From there, the water would flow by gravity to the western states. NAWPA would initially provide 137.5 billion m^3 of water annually to seven provinces of Canada, 33 states in the USA and three northern states of Mexico. The total power generation would be 100 million k W/yr. Out of this, 30 million k W/yr would be utilised by the pumping requirements of the project. NAWPA is a gigantic project and its environmental and social costs have yet to be comprehensively determined. Its international and political implications are quite enormous and have yet to be fully realised.

Much of the excess water will originate from Canada and will be used in the United States. Canadian public opinion is solidly against the transfer of such a massive quantity of water to its neighbouring economic giant to the south, the United States. The Canadians fear that once such developments have taken place in the United States to use this excess water, it would not be politically and economically possible to stop the export of water, even after the treaty period of whatever duration is over. They are concerned that in the future, when more water will be necessary for Canada's own use and further development, it would no longer be available.

Even if this major international problem was solved in favour of water transfer, and it was further clearly shown that technical, economic, social and environmental problems could be overcome (these have yet to be resolved and in most cases have not even been identified), current experiences indicate that the litigations alone would postpone the implementation of such a project for at least a decade, perhaps even longer. In other words, if present trends continue, it is highly unlikely that construction of such a project would start before the end of the century at the earliest-if it is implemented at all!

Problems in Large-scale Water Transfer

There are a number of problems associated with large-scale transfer of water from one region to another. The magnitude of the problems will differ from one project to another, but some of the major variables that should be considered are the following:

Physical System

(a) Water Quantity: level; discharge; velocity; groundwater; losses.

(b) Water Quality: sediments; nutrients; turbidity; salinity and alkalinity; temperature effects; toxic chemicals.

(c) Land Implications: erosion; sedimentation; salinity; alkalinity; waterlogging; changes in land use patterns; changes in mineral and nutrient contents of soil; earthquake inducement; other hydrogeological factors.

(d) Atmosphere: temperature; evapotranspiration; changes in microclimate; changes in macroclimate.

Biological System

(a) Aquatic: benthos; aufwuchs; zooplankton; phytoplankton; fish and aquatic vertebrates; plants; disease vectors.

(b) Land-based: animals; vegetation; loss of habitat; enhancement of habitat.

Human System

(a) Production: agriculture; aquaculture; hydropower; transportation (navigation); manufacturing; recreation; mining.

(b) Socio-cultural: social costs, including resettlement of people; infrastructural developments; anthropological effects; political implications.

In the past, long-distance mass transfer of water has often proved to be an emotive issue. Every time such a project is proposed, public controversies have become the rule rather than the exception. The proponents argue that their technical excellence, economic benefits and overall contributions to development make them indispensable to society. The opponents on the other hand point out that their total social and environmental costs are far too high and hence unacceptable to society. Thus, in all likelihood, one segment of society will lobby for such a development while another section will oppose the same project for different reasons.

To a certain extent, such conflicts can be explained by analysing the nature of the beneficiaries, which is seldom carried out for any water resources development project. It is inevitable that any development project will benefit some citizens more than others and, frequently, some citizens may have to bear additional costs, either tangible (i.e. heavier tax liabilities) or intangible (i.e. social and environmental). The proponents are often the net beneficiaries and the opponents tend to be those who pay the costs.

Even scientists are strongly divided on the question of such transfers. For example, Howe and Easter have concluded that such water transfers are likely to prove expensive to nations, except under certain "rescue operation" type of cases. But Wells has pointed out that water imports to the high plains of Texas are not only economically feasible but also that the state simply cannot afford not to import water to that area. If many of the past and present experiences on long-distance water transfer are reviewed critically, the following major issues emerge:

(1) Mass transfer of water is often justified by considering only the direct cost of transporting water. Seldom are the values of services foregone by the exporting region due to reduction of their water availability, i.e. the opportunity costs of exported water analysed.

(2) Various other feasible alternatives to interbasin water transfer are often not investigated. There is a tendency within the engineering and economic

professions to opt for technological solutions-"soft" options tend to be neglected. Since water resources development is dominated by these two professions, there is a tendency to opt for technological fixes before all viable alternatives are explored. Among possible options are:

— more efficient use of available water;

— re-use of waste water ;

— better management of watersheds;

— improved integration of surface and groundwater supplies;

— changing cropping patterns.

(3) The agricultural sector is usually the major beneficiary of water transfer projects. Thus, much of the analysis often concentrates on agricultural benefits, and social objectives like income redistribution, alteration in regional growth rates and patterns, reduction in unemployment and environmental protection are not considered. To a certain extent this can be explained by the difficulties encountered in quantifying social and environmental benefits and costs.

(4) Opposition to large-scale mass transfer of water in developed countries, especially for interstate and international projects, is likely to increase as more and more water is required for various purposes. Transfer of water from one country to another, or even between states within a country, can seldom be achieved without controversy. Technical and economic feasibility studies are comparatively easy to conduct; the real problem lies in their public and social acceptability. Considered purely from a rational viewpoint, such reasoning is hard to

explain since states and countries freely export other resources like minerals, hydrocarbons and agricultural products. Public sentiment, for whatever reasons, often seems to be strongly against water exports and this concern, not surprisingly, is reflected in the political process. This may perhaps be explained by the "water is different" syndrome and in all probability is unlikely to change in the near future.

(5) The legal implications of interstate and international water transfers are quite complicated. Adequate legal and institutional frameworks for such developments are rare, and currently there is no process for speedy resolution of these conflicts. This point can be quickly realised by the present disputes that exist between states and between nations on the use of interstate and international rivers and lakes.

(6) Since the mid-sixties, opposition to mass transfer of water has increased significantly on environmental and social grounds in developed countries. Many segments of society are no longer willing to accept such social and environmental costs as the price of progress.

13

Water Quality Monitoring

Qualitative and quantitative measurements are needed from time to time to constantly monitor the quality of water from the various sources of supply. The harbour-master should then ensure appropriate water treatment within the fishery harbour complex as well as initiate remedial measures with the suppliers when water supply from outside is polluted. Water sampling and analysis should be done by ISO-certified laboratories. Wherever laboratories available locally are not ISO-certified, it is advisable to get their quality assessed by an ISO-certified laboratory by carrying out collaborative tests to ensure that variation in the accuracy of results is sufficiently small.

Unreliable results exacerbate problems of pollution when corrective action cannot be taken in time. Sampling and monitoring tests should be carried out by qualified technicians. Depending on the actual state of the fishing harbour infrastructure and environmental conditions in and around the harbour, monitoring should be carried out according to a specific programme for each source of water supply.

Borewells

Contamination may arise from pollutants entering the water table some distance from the port or from sewage entering the borehole itself in the port area through cracked or corroded casings.

Municipal mains: Supply could be contaminated at source or through corroded pipelines leading to the fishery harbour. Mixing with sewage lines due to defective piping has been known to occur often. Complete tests should be carried out every half year, and the authorities should be informed when results indicate contamination.

Water tanks and reservoirs: Both types of structure are prone to bacterial growth if the residual chlorine levels in them are low or non-existent. Testing may not be necessary if periodic scrubbing is carried out. Bacteriological tests should be done at least half-yearly.

Harbour basin water: Typically, harbour basins are tested yearly. However, in areas where monsoons are very active, it may be advisable to test at the peak of the dry season when effluent point discharges tend to remain concentrated in the water body and again during the wet season when agriculture run-off may be considerable. Another critical period for harbours is the peak of the fishing season when the harbour is at its busiest and vessel-generated pollution is likely to be at its peak.

Water Quality Tests

Testing procedures and parameters may be grouped into physical, chemical, bacteriological and microscopic categories.

— Physical tests indicate properties detectable by the senses.

— Chemical tests determine the amounts of mineral and organic substances that affect water quality.

— Bacteriological tests show the presence of bacteria, characteristic of faecal pollution.

Physical tests

Colour in water may be caused by the presence of minerals such as iron and manganese or by substances of vegetable origin such as algae and weeds. Colour tests indicate the efficacy of the water treatment system.

Turbidity in water is because of suspended solids and colloidal matter. It may be due to eroded soil caused by dredging or due to the growth of microorganisms. High turbidity makes filtration expensive. If sewage solids are present, pathogens may be encased in the particles and escape the action of chlorine during disinfection.

Odour and taste are associated with the presence of living microscopic organisms; or decaying organic matter including weeds, algae; or industrial wastes containing ammonia, phenols, halogens, hydrocarbons. This taste is imparted to fish, rendering them unpalatable. While chlorination dilutes odour and taste caused by some contaminants, it generates a foul odour itself when added to waters polluted with detergents, algae and some other wastes.

Chemical Tests

pH is a measure of hydrogen ion concentration. It is an indicator of relative acidity or alkalinity of water. Values of 9.5 and above indicate high alkalinity while values of 3 and below indicate acidity. Low pH values help in effective chlorination but cause problems with corrosion. Values below 4 generally do not support living organisms in the marine environment. Drinking water should have a pH between 6.5 and 8.5. Harbour basin water can vary between 6 and 9. B.O.D. denotes the amount of oxygen needed by microorganisms for stabilisation of decomposable organic matter under aerobic conditions. High B.O.D. means that there is less of oxygen to support life and indicates organic pollution.

Bacteriological tests

For technical and economic reasons, analytical procedures for the detection of harmful organisms are impractical for routine water quality surveillance. It must be appreciated that all that bacteriological analysis can prove is that, at the time of examination, contamination or bacteria indicative of faecal pollution, could or could not be demonstrated in a given sample of water using specified culture methods. In addition, the results of routine bacteriological examination must always be interpreted in the light of a thorough knowledge of the water supplies, including their source, treatment, and distribution. Whenever changes in conditions lead to deterioration in the quality of the water supplied, or even if they should suggest an increased possibility of contamination, the frequency of bacteriological examination should be increased, so that a series of samples from well chosen locations may identify the hazard and allow remedial action to be taken.

Whenever a sanitary survey, including visual inspection, indicates that a water supply is obviously subject to pollution, remedial action must be taken, irrespective of the results of bacteriological examination. For unpiped rural supplies, sanitary surveys may often be the only form of examination that can be undertaken regularly. The recognition that microbial infections can be waterborne has led to the development of methods for routine examination to ensure that water intended for human consumption is free from excremental pollution. Although it is now possible to detect the presence of many pathogens in water, the methods of isolation and enumeration are often complex and time-consuming. It is therefore impractical to monitor drinking water for every possible microbial pathogen that might occur with contamination.

A more logical approach is the detection of organisms normally present in the faeces of man and other warm-blooded animals as indicators of excremental pollution, as well as of the efficacy of water treatment and disinfection. The presence of such organisms indicates the presence of faecal material and thus of intestinal pathogens. (The intestinal tract of man contains countless rod-shaped bacteria known as coliform organisms and each person discharges from 100 to 400 billion coliform organisms per day in addition to other kinds of bacteria). Conversely, the absence of faecal commensal organisms indicates that pathogens are probably also absent. Search for such indicators of faecal pollution thus provides a means of quality control.

The use of normal intestinal organisms as indicators of faecal pollution rather than the pathogens themselves is a universally accepted principle for monitoring and assessing the microbial safety of water supplies. Ideally, the finding of such indicator bacteria should denote the possible presence of all relevant pathogens. Indicator organisms should be abundant in excrement but absent, or present only in small numbers, in other sources; they should be easily isolated, identified and enumerated and should be unable to grow in water. They should also survive longer than pathogens in water and be more resistant to disinfectants, such as chlorine. In practice, these criteria cannot all be met by any one organism, although many of them are fulfilled by coliform organisms, especially Escherichia coli as the essential indicator of pollution by faecal material of human or animal origin.

Treatment Methods

Treatment of raw water to produce water of potable quality can be expensive. It is advisable to determine the quantity of water needing treatment, as not all water used in a fishery harbour or processing plant needs to be of potable quality. Sizing of the equipment is crucial to produce acceptable water at reasonable cost. The main point to remember is that separate systems and pipelines are required for potable and non-potable water to avoid cross contamination.

Each system must be clearly identified by contrasting coloured pipelines. Water used for drinking, cleaning fish and ice-making must be free from pathogenic bacteria and may require secondary treatment or even complete treatment depending on chemical elements that need to be removed. Water for other needs like general cleaning may perhaps need only primary treatment.

Primary Treatment

There are four methods of primary treatment: chlorination; ozone treatment; ultraviolet treatment; and membrane filtration.

Chlorination: Fresh or sea water can be chlorinated using either chlorine gas or hypochlorites. Chlorinated water minimises slime development on working surfaces and helps control odour.

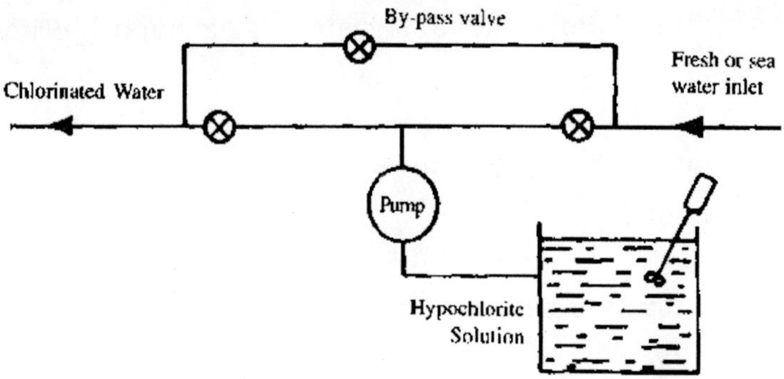

Figure 1. Chlorination Treatment

The main advantages of using chlorine gas are:

— It is the most efficient method of making free chlorine available to raw water.

— It lowers the pH of the water slightly.

— Control is simple; testing simple; and it is not an expensive method.

The main disadvantages are:

— Chlorine gas is toxic and can combine with other chemicals to form combustible and explosive materials.

— Automatic control systems are expensive.

— Chlorine cylinders may not be readily available at small centres.

— Chlorine expands rapidly on heating and hence the cylinders must have fusible plugs set at 70°C. It also reacts with water, releasing heat. Water should not therefore be sprayed on a leaking cylinder.

Hypochlorites are generally available in two forms—sodium hypochlorite solution normally available at 10%

concentration and calcium hypochlorite available as a powder. The main disadvantages of using hypochlorites are:

- — Calcium hypochlorite is not stable and must be stored in air-tight drums.
- — Sodium hypochlorite is quite corrosive and cannot be stored in metal containers
- — Sodium hypochlorite must be stored in light proof containers.
- — It is difficult to control the rate of addition of hypochlorites in proportion to water flow.
- — Hypochlorites raise the pH in water.
- — They are more expensive than chlorine gas.

It is important to understand the manner in which chlorine or chlorine-releasing substances behave when added to water, depending on other substances present.

- — When water contains reducing substances like ferrous salts or hydrogen sulphide, these will reduce part of the added chlorine to chloride ions.
- — When water contains ammonia, organic matter, bacteria and other substances capable of reacting with chlorine, the level of free chlorine will be reduced.
- — If the quantity of chlorine added is sufficiently large to ensure that it is not all reduced or combined, a portion of it will remain free in the water. This is termed as residual free chlorine or free chlorine.

When chlorine reacts chemically as in the first two cases, it loses its oxidising power and consequently its disinfecting properties. Some ammoniacal chlorides however still retain some disinfecting properties.

Chlorine present in this form is termed residual combined chlorine or combined chlorine. From the standpoint of disinfection, the most important form is free chlorine. Routine analysis always aims at determining at least the free chlorine level.

Ozone treatment: Though the principle is relatively simple, this method needs special equipment, supply of pure oxygen and trained operators. Ozone is generated by passing pure oxygen through an ozone generator. It is then bubbled through a gas diffuser at the bottom of an absorption column, in a direction opposite to the flow of raw water. Retention or contact time is critical and the size of the absorption column depends on the water flow.

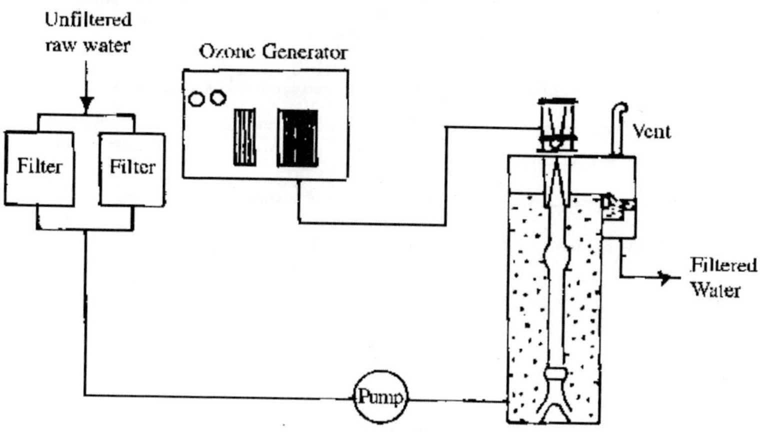

Figure 2. Ozone Treatment

The main advantages of ozone treatment are:

— Ozone is a much more powerful germicide than chlorine especially for faecal bacteria.

— It reduces turbidity of water by breaking down organic constituents.

— The process is easily controlled.

The disadvantages are:

— Pure oxygen may not be readily available locally.

— Ozonized water is corrosive to metal piping.

— Ozone decomposes rapidly into oxygen.

— Water has to be aerated prior to use to remove the ozone.

Ultraviolet irradiation treatment: This method is often used to treat drinking water. Successful commercial installations have been made to purify sea water in large fish processing plants.

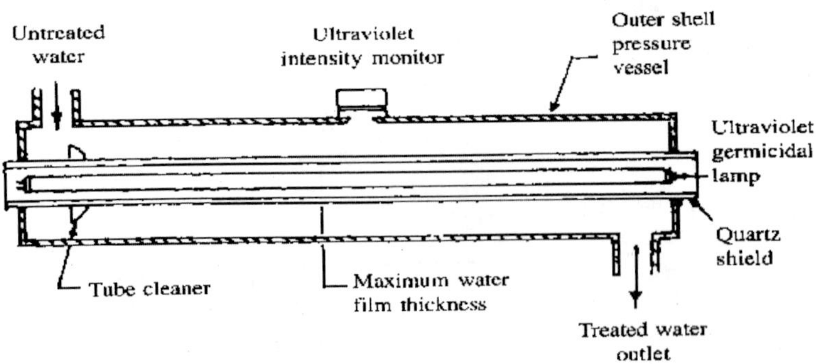

Figure 3. Ultraviolet Irradiation Treatment

The main advantages of U-V treatment are:

— U-V rays in the range of 2500-2600 Angstrom units are lethal to all types of bacteria.

— There is no organoleptic, chemical or physical change to the water quality.

— Overexposure does not have any ill effects.

The main disadvantages are:

— Electricity supply should be reliable.

— Turbidity reduces efficiency.

— Water may require prior treatment like filtration.

— The unit requires regular inspection and maintenance.

— Thickness of the water film should not exceed 7.5 cm.

Membrane filtration: Osmotic membrane treatment methods are generally expensive for commercial scale installations. Combinations of membrane treatment with U-V treatment units are available for domestic use.

Secondary Treatment

Secondary treatment of water consists of sedimentation and filtration followed by chlorination. Sedimentation can be carried out by holding the raw water in ponds or tanks. The four basic types of filtration are cartridge filtration, rapid sand filtration, multimedia sand filtration, and up-flow filtration.

Cartridge filtration: This system is designed to handle waters of low turbidity and will remove solids in the 5 to 100 micron range. The main advantages are:

— Low cost and 'in-line' installation.

— Change of cartridge is simple.

— Operation is fool-proof. Once the cartridge is clogged, flow simply stops.

The main disadvantages are:

— Sudden increase in turbidity overloads the system.

— Cartridges may not be readily available and large stocks may be required.

Rapid sand filtration: This system consists of a layer of gravel with layers of sand of decreasing coarseness above the gravel. As solids build up on top, flow decreases until it stops. This is corrected by back-flushing the system to remove the solid build up on top, Figure 4.

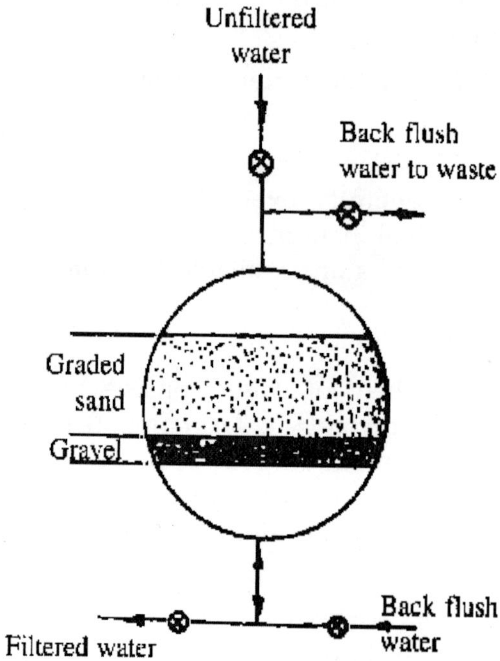

Figure 4. Rapid Sand Filtration

The main advantages are:

— Cost of filtration media is negligible.

— Operation is simple.

The main disadvantages are:

— A holding tank for filtered water is required to provide clear water back flushing.

— Pumping loads increase as sediments build up.

Multimedia sand filtration: This system is similar to the rapid sand filtration method.

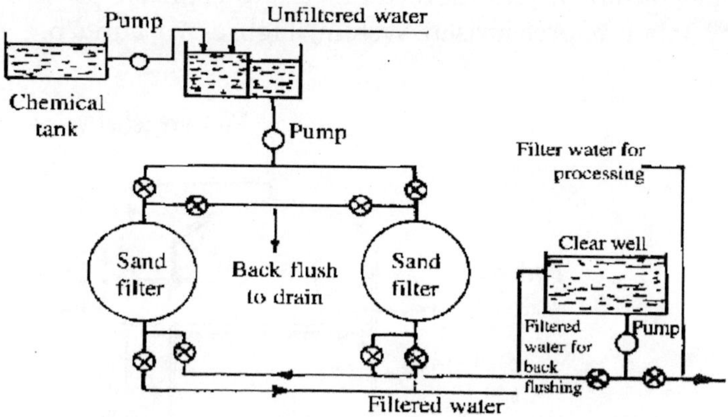

Figure 5. Conventional Sand Filtration

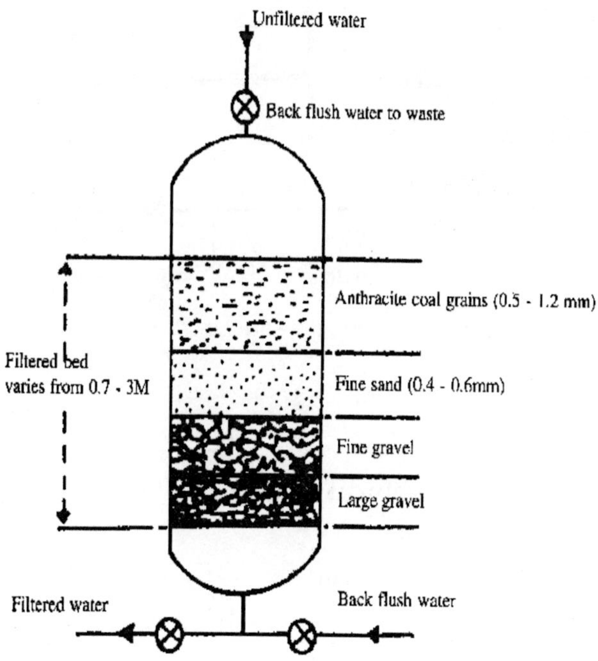

Figure 6. Multumedia Sand Filtration

Up-flow filtration: Filtration can be at atmospheric pressure or by using a pressurised system, Figures 7a and 7b.

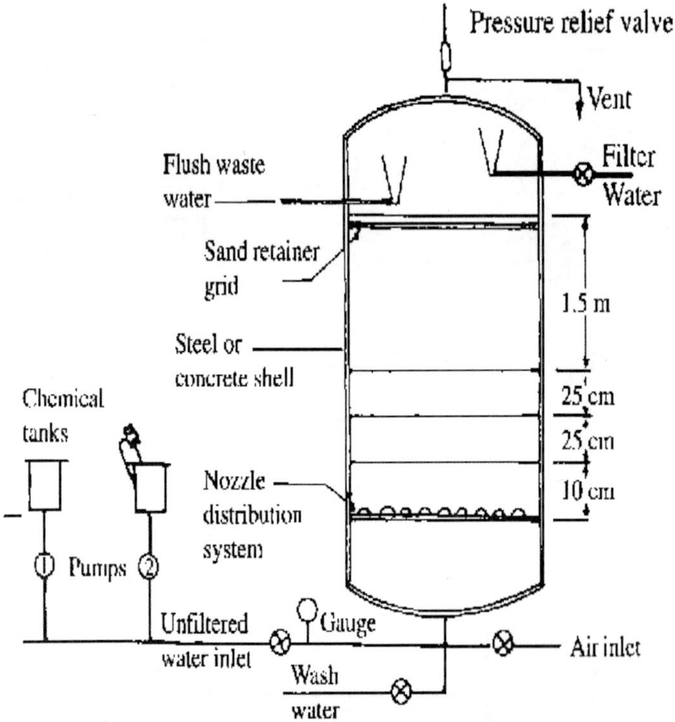

Figure 7a. Automatic Pressure Up-flow Filter

The main advantages are:

— High flow rates are easily attained.

— Water with turbidity up to 1500 ppm can be handled.

— Degree of filtration can be easily adjusted.

— The filter bed can be easily cleaned using the filtered water.

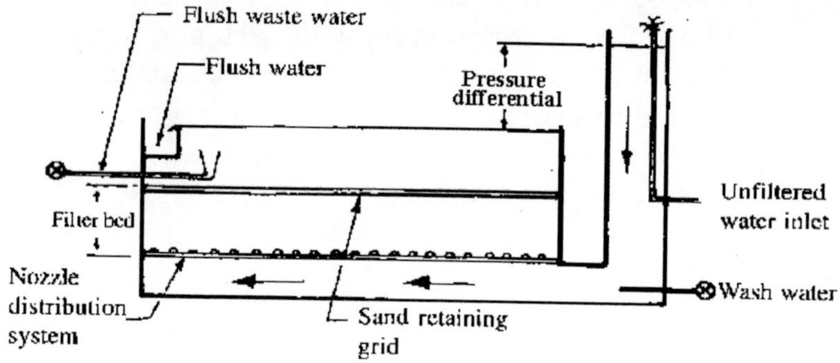

Figure 7b. Pressure Type Up-flow Filter

The main disadvantage is:

— Close supervision is necessary to ensure that the filter bed does not rupture.

Complete Treatment

Complete treatment consists of flocculation, coagulation, sedimentation and filtration followed by disinfection. Flocculation and coagulation will assist in removing contaminants in the water, causing turbidity, colour odour and taste which cannot be removed by sedimentation alone. This can be achieved by the addition of lime to make the water slightly alkaline, followed by the addition of coagulants like Alum (aluminium sulphate), ferric sulphate or ferric chloride.

The resultant precipitate can be removed by sedimentation and filtration. Chemical treatment may be required to reduce excessive levels of iron, manganese, chalk, and organic matter. Such treatment is usually followed by clarification. Iron may be removed by aeration or chlorination to produce a flocculant which can be removed by filtration. Manganese may be removed by aeration followed by adjustment of pH and up-flow filtration. Most colours can be removed by treatment with ferric sulphate to precipitate the colours.

Bibliography

Ayers R.S. and Westcot D.W., "Water quality for agriculture", *FAO Irrigation and Drainage Paper* 29, FAO, Rome. 97 p., 1976

Beltran J.M., "Drainage and reclamation of salt-affected soils, Bardenas Area, Spain", *Publication 24, International Institute for Land Reclamation and Improvement (ILRI)*, Wageningen, The Netherlands, 1978

Bernstein L., "Salt tolerance of plants", *USDA Agricultural Information Bulletin*, 1964

Bingham F.T., Mahler R.J. and Sposito G., "Effects of irrigation water composition on exchangeable sodium status of a field soil", *Soil Science*, 1979

Bucks D.A., Nakayama F.S. and Gilbert R.G., "Trickle irrigation water quality and preventive maintenance", *Agricultural Water Management*, 1979

Chapman H.D. and Pratt P.F., "Methods of analysis for soils, plants and waters", University of California, Division of Agricultural Science, 1961

Dewan H.C., et al. "Irrigation water quality in the Yemen Arab Republic", *Report by Soil and Water Research Section*, Central Agricultural Research Organization, 1978

Dewis J. and Freitas F., "Physical and chemical methods of soil and water analysis", *FAO Soils Bulletin* 10. FAO, Rome, 1970

Dieleman P.J. and Trafford B.D., "Drainage testing", *FAO Irrigation and Drainage Paper* 28. FAO, Rome, 1976.

Dieleman P.L. et al., "Drainage design factors - An Expert Consultation on Drainage Design Factors", 22–29 October 1979, *FAO Irrigation and Drainage Paper* 38. FAO, Rome, 1980.

Doneen L.D., "Water quality requirements for agriculture", *National Symposium on Quality Standards for National Waters*, Educational Series No. 161:213–218. University of Michigan, Ann Arbor, 1967

Eckenfelder W.W., "Principles of water quality management", CBI Pub. Co., Massachusetts, 1980

FAO/Unesco, "Irrigation, Drainage and Salinity", *An International Sourcebook*, Paris, Unesco/Hutchinson (Publishers), London, 1973.

Ford H.W. and Tucker D.P.H., "Water quality measurements for drip irrigation systems", *Florida Agricultural Experiment Stations Journal Series* No. 5598. Lake Alfred, Florida, 1974

McJunkin F.E. et al., "Water and human health", *US Agency for International Development*, Department of State, Washington DC, 1982

Rhoades J.D., "Drainage for salinity control", *In: Drainage for Agriculture*. Van Schilfgaarde J. (ed). Amer. Soc. Agron. Monograph No. 17, 1974

______________ ., "New strategy for using saline waters for irrigation", *In: Water Today and Tomorrow*, Proc. Speciality Conference at Flagstaff Arizona. ASCE, 1984.

Shalhavet J., "Management of irrigation with brackish water", *In: Soil Salinity under Irrigation*, I. Shainberg and J. Shalhavet (eds). Ecological Studies 51:298–318. Springer-Verlag, New York, 1984

Index